Richard Chandler Alexander Prior

On the Popular Names of British Plants

Richard Chandler Alexander Prior

On the Popular Names of British Plants

ISBN/EAN: 9783744674164

Printed in Europe, USA, Canada, Australia, Japan

Cover: Foto ©berggeist007 / pixelio.de

More available books at **www.hansebooks.com**

POPULAR NAMES OF BRITISH PLANTS.

ON THE

POPULAR NAMES

OF

BRITISH PLANTS,

BEING AN EXPLANATION OF THE

ORIGIN AND MEANING

OF THE NAMES OF OUR

INDIGENOUS AND MOST COMMONLY CULTIVATED SPECIES.

BY

R. C. A. PRIOR, M.D.,

FELLOW OF THE ROYAL COLLEGE OF PHYSICIANS OF LONDON, AND OF THE LINNEAN
AND OTHER SOCIETIES;
TRANSLATOR OF "ANCIENT DANISH BALLADS."

SECOND EDITION

E PUR SI MUOVE

WILLIAMS AND NORGATE,

14, HENRIETTA STREET, COVENT GARDEN, LONDON.
AND 20, SOUTH FREDERICK STREET, EDINBURGH.
1870.

HERTFORD :

PRINTED BY STEPHEN AUSTIN.

ADVERTISEMENT.

In publishing a second edition of this work, I beg to express my warmest thanks and my obligation to many friends, and to many correspondents not personally known to me, but more particularly to Mr. Leo Grindon, of Manchester, and Mr. M. P. Edgworth, and the Rev. J. C. Atkinson, for much valuable information that they have given me, and for the correction of several oversights and errors.

R. C. A. P.

HALSE HOUSE, NEAR TAUNTON,
Nov. 1, 1870.

INTRODUCTION.

THE authors of our several Floras, and other systematic writers, have been careful to translate the Greek and Latin names of our plants, and, as far as it is known, to explain their meaning, but have passed over the popular ones, as though the derivation of these were too obvious to require any notice. This is far indeed from being the case. Our excellent lexicons and Latin dictionaries enable us in most cases to understand the former with comparative facility, but in the very backward state of English etymology, as exhibited in books of reference, it is impossible, without a great waste of time and trouble, to discover the real meaning of the latter; of those more particularly which date from an early period. It is the object of the following Vocabulary to supply the defect.

The Anglo-Saxon names, during the period of nearly five hundred years that intervened between the Conquest and the revival of botanical inquiry in the sixteenth century, had, the most of them, fallen into disuse, and been replaced with others taken from Latin and French, or transferred to plants to which they did not originally belong. They have probably been nearly all of them preserved to us in ancient manuscripts; but it is difficult to ascertain what were the several plants that were meant by them. Indeed it is not likely that in earlier times any great number of our indigenous species had been carefully distinguished. It is only when nations have arrived at a high state of culture, that they are curious about objects of Natural History, as such, or have special names for any but a few of the more conspicuously useful, beautiful, or troublesome of them. Our fruit and timber trees, the cereal grains, and several potherbs and medicinal plants, have the same at the present day as they bore a thousand years ago; but by far the greater number of our other species have only such as have been given to them within the last three hundred years. These, for the most part, were introduced from abroad; for in the accurate study of living plants the continental nations took the lead, and our own early herbalists did little more than ascertain which they meant, and apply their names to our own.

In the selection of these the father of English Botany, Dr. William Turner, set his successors a laudable example by keeping as closely as possible to the Flemish and German, as languages more akin to our mother-tongue, and intelligible to the uneducated, than Greek and Latin. Lyte in his excellent translation of Dodoens did the same, and was worthily followed by Gerarde, and by Parkinson. The works of later herbalists are little else than transcripts of what was published by these four. Turner's Herbal came out in three parts between 1551 and 1568; Lyte's in 1578; Gerarde's in 1597; a new edition of it by T. Johnston in 1632; and Parkinson's two works, his Paradisus Terrestris, and his Theatre of Botany, in 1629 and 1643. The Grete Herball, the Little Herbals, and Macer's Herbal, Batman's Bartholomew de Glantvilla, and some other black-letter books of an earlier date than Turner's, are of scarcely any assistance to us, from the difficulty there is to discover by their very inadequate descriptions, what plants they mean. The ancient vocabularies published by Jos. Mayer and Wright, and by Halliwell and Wright, and others in the British Museum* and foreign libraries, are, for the same reason, very seldom available. Some very valuable manuscripts have, since the first edition of this work, been published, with a translation and glossary, by the Rev. Oswald Cockayne, in his " Anglo-Saxon Leech-doms," a work of great interest, and one to which reference is made in the following pages.

There are distinguished botanists at the present day who look upon popular names as leading to confusion, and a nuisance, and who would gladly abandon them, and ignore their existence. But this is surely a mistake, for there will always be ladies and others, who, with the greatest zeal for the pursuit of Natural History, have not had the

---

* I have here the agreeable duty of acknowledging the kindness of Mr. J. J. Bennett, the Curator of the Botanical collections, in most handsomely placing at my disposal many extracts from these manuscripts, that he had made for a similar undertaking.

opportunity of learning Greek and Latin, or have forgotten
it, and who will prefer to call a plant by a name that they
can pronounce and recollect. We need but to ask ourselves, what success would have attended the exertions
of the late excellent and benevolent Professor Henslow
among the little pupils of his village school, if he had
used any names but the popular ones.

Besides, admitting to the full all that can be urged
against them from a purely botanical point of view,
we still may derive both pleasure and instruction from
tracing them back to their origin, and reading in
them the habits and opinions of former ages. In
following up such an analysis we soon find that we are
travelling far away from the humble occupation of the
herbalist, and are entering upon a higher region of literature, the history of man's progress, and the gradual
development of his civilization. Some of the plants that
were familiar to our ancestors in Central Asia, bear with
us to this day the very names they bore there, and as
distinctly intimate by them the uses to which they were
applied, and the degree of culture which prevailed where
they were given, as do those of the domestic affinities the
various occupations of the primeval family. The names of
animals, with which many are compounded, carry us still
further back, or to still more distant regions; for in some
cases it is impossible now to deduce any meaning from
them at all, and it is probable that these names may have
been adopted, with the knowledge of the animal, from an

entirely alien nation. In such, for instance, as *hound* and *ox*, we have unquestionable proof that they must have been given to those animals, before the existing dialects of our ancient mother-tongue had assumed their distinctive form; and this must have been at an immensely remote point of time. For to educe from the same language others so different from one another, not only in their vocabulary, but in their grammatical construction and declensions, as were already in their earliest known state the oldest of them with which we are acquainted, the Sanskrit, Latin, Greek, and Gothic, required a period, not of centuries merely, but millennia.

The most interesting, in this respect, of the names that have come down to us, are those which date from a period antecedent to the settlement of the German race in England, names which are deducible from Anglo-Saxon roots, and identical, with allowance for dialectic peculiarities, in all the High and Low German, and Scandinavian languages, and, what is particularly worthy of our attention, each of them expressive of some distinct meaning. These will prove, what with many readers is a fact ascertained upon other evidence, such as the contents of sepulchral mounds, traditionary laws, and various parallel researches, that the tribes which descended upon Britain had entered Europe, not as a set of savages, or wandering pastoral tribes, or mere pirates and warriors, but as colonists, who, rude as they may have been in dress and manners, yet, in essential points, were already a civilized people. It will

be seen at the same time that they must have come from
a colder country; for while these names comprehend the
Oak, Beech, Birch, Hawthorn, and Sloe, trees that extend
far into Northern Asia, they do not comprise the Elm,
Chesnut, Maple, Walnut, Sycamore, Holly, or any evergreen, except of the fir tribe, or Plum, Pear, Peach, or
Cherry, or any other fruit-tree, except the Apple.   For all
these latter they adopted Latin names, a proof that at the
time when they first came into contact with the Roman
provincials on the Lower Rhine, they were not the settled
inhabitants of the country they were then occupying, but
foreigners newly arrived there as colonists or conquerors
from a country where those trees were unknown.   It is
remarkable that the early Greek writers make no mention
of any German tribes, but represent the Scythians as the
next neighbours of the Celts, and this difference in the
names of the one set of trees and the other, and the names
which they adopted being Roman, and not Celtic, suggests
that the Germans had come down from the north-east not
very long before the Christian era, and intruded themselves, as a wedge, between those two more anciently
recorded nations.

   There seems to be much misapprehension in respect to
this great movement of the Eastern races which broke up
the Roman empire.   The subject is one, into which it
would here be out of place to enter fully, and it has been
largely treated by J. Grimm in his admirable Geschichte
der Deutschen Sprache.   But even in the following voca-

bulary we shall see evidence of the continuous advance of a civilized race from the confines of India to these islands, and nothing indicative of a great rush from the North of wild hordes bent upon robbery and destruction, as it has been usually represented to have been. The gradual drying of the Caspian Sea left the interior of Asia more and more barren, the knowledge of the useful metals facilitated the conquest of the savages of the West, and it is likely that predatory bands of Huns and Turks and allied nomadic nations accelerated the movement by rendering the labours of agriculture less remunerative. Thus the migration, being one that proceeded from constantly acting causes, extended over many centuries. Let us lay aside all prepossessions, and inquire what light is thrown by the following vocabulary upon the real state of the Germanic tribes at that period.

In these mere names of plants, setting aside all other sources of information, we discover that these people came from their home in the East with a knowledge of letters, and the useful metals, and with nearly all the domestic animals; that they cultivated oats, barley, wheat, rye, and beans; built houses of timber, and thatched them; and, what is important, as showing that their pasture and arable land was intermixed, and acknowledged as private property, they hedged their fields and fenced their gardens. Cæsar denies this; but the frontier tribes, with whom he was acquainted, were living under certain peculiar Mark laws, and were, in fact, little else than an army on its march. The unquestionably native, and not Latin or Celtic origin of such names as Beech and Hawthorn, of Oats and Wheat, prove that although our ancestors may have been indebted to the provincials of the empire for their fruit-trees, and some other luxuries, for a knowledge of the fine arts, and the Latin literature, and a debased Christianity, the more essential acquirements upon which their prosperity and progress as a nation depended were already in their possession. Like the scattered lights that a traveller from the wilderness sees here and there in a town that lies shrouded in the darkness of night in a valley beneath him, and the occasional indistinct and solitary voice of some domestic animal, that for a moment breaks the silence, these distant echoes of the past, these specks that glimmer from its obscurity, faint as they are, and few and far between, assure us that we are contemplating a scene of human industry, and peace, and civilization.

In this respect the inquiry is one of the highest interest. In another it is probable that some who consult these pages will be disappointed. The names have usually been given to the plants from some use to which they were applied, and very few of them bear any trace of poetry or romance. In short, our Sweet Alisons and Herb Truloves, our Heartseases, Sweet Cicelies, and Sweet Williams resolve themselves into sadly matter-of-fact terms, which arose from causes very different from the pretty thoughts with which they are now associated, and sometimes, as in the case of the Forget-me-not, were suggestive of very disagreeable qualities. In many cases, as in that of the hawkweed, the miltwaste, and the celandine, they refer to virtues that were ascribed to the plants from the use that birds and other animals were supposed to make of them. Many more have been given to them in accordance with the so-called *doctrine of signatures.* This was a system for discovering the medicinal uses of a plant from something in its external appearance that resembled the disease it would cure, and proceeded upon the belief that God had in this manner indicated its especial virtues. Thus the hard stony seeds of the Gromwell must be good for gravel, and the knotty tubers of scrophularia for scrofulous glands ; while the scaly pappus of Scabiosa showed it to be a specific in leprous diseases, the spotted leaves of Pulmonaria, that it was a sovereign remedy for tuberculous lungs, and the growth of Saxifrage in the fissures of rocks, that it would disintegrate stone in the bladder. For, as Wm. Coles tells us in his Art of Simpling, ch. xxvii : " Though Sin and Sathan have plunged mankinde into an Ocean of Infirmities, yet the mercy of God which is over all his workes, maketh Grasse to grow upon the Mountaines, and Herbes for the use of men, and hath not only stamped upon them a distinct forme, but also given them particular Signatures, whereby a man may read, even in legible characters, the use of them."

Other names we shall find relate to the economical uses

to which the plants were once applied. Some few are descriptive; some refer to the legends or the ceremonies of the Roman Catholic Church; some to the elegant mythology of the Greeks; some to a vulgar joke. In thinking over these names, and the antiquated notions that they represent, we are led at every moment to recall the times from which they date, to picture to ourselves the living figures of our ancestors, to hear them speaking their obsolete dialect, and almost to make the weeds that shadow their grave tell more than their tombstone of its sleeping inhabitants.

The terms with which we have to deal may for convenience be referred to two groups, as Germanic, or Romanic. To the former belong such as are of Anglo-Saxon, German, or Low German, or Scandinavian origin, and to the latter such as are French, or derived from other forms of debased Latin, including a few adopted into it from the Arabic. When a word falls within the first group, we find great assistance in Dr. Bosworth's and J. Jamieson's Dictionaries, and in the works of Adelung, Bopp, Pott, Diefenbach, and the brothers Grimm, and in those of the Frisian and Scandinavian writers. French words, from the loss of those Celtic dialects with which the Latin element of the language was corrupted, and the extreme degree of debasement to which it has arrived, are of much more difficult analysis. For these we have the assistance of Diez's Wörterbuch der romanischen Sprachen, and Scheler's Dictionnaire d'Etymologie Française, and the admirable Dictionnaire etymologique of Emile Egger, and the copious Dictionnaire de la Langue Française of Littré. A large number of the names referrible to this group have been adopted from the Latin of the Middle Ages, a jargon that, with many peculiarities in each country, was at one period used all over the West of Europe, and is explained in the great Lexicon Mediæ Latinitatis of Ducange. These names, obscure as they often were from the first, have been so corrupted by ignorant copyists as in many cases to defy all analysis, and render it necessary to refer to old vocabularies, catalogues, and herbals to discover their meaning. We might have expected many to have been derived from the language of the ancient Britons; but, as far as I am aware, "Maple" is the only one; and there are very few indeed that have been adopted from the modern Welsh, or from the Erse or Gaelic.

As the term "Ind-European" will be frequently used, and some may refer to the following vocabulary who have not entered into philological speculations of this kind, it is necessary to mention that the analysis of words, and the comparison of their roots and grammatical structure, have proved that all the principal languages of alphabetical literature, exclusive of the Arabic and its allies, are intimately connected with the ancient dialects of Persia and Northern India. This has been considered by many writers as a proof that all the nations which speak them have descended from a common stock, and although this inference as to the people may be incorrect, still, in a philological point of view, we may treat the languages as sister descendants of some Asiatic parent which has long since perished, and rank with them such other dialects as agree with these in their roots and structure. Under "Ind-European," then, will be comprised Sanskrit and Zend, and all the Indian and Persian dialects that are related to them; Greek; Latin, and its modern varieties; Celtic; Gothic, and all the other Germanic and Scandinavian dialects; Lithuanian, and Slavonian; but not Basque, Lapp, Finn, Magyar, or Turkish. The language of the Indian Vedas, as the oldest existing member of the family, is that to which linguists refer in searching for the roots of words of this class, itself no more than the representative of another still more ancient one, which is utterly lost.

In order to avoid a long word, and for no other apparent reason, it has been proposed of late to supersede the very expressive and most unobjectionable term of "Indo-"* or "Ind-European," and to substitute for it that of "Aryan," which it is to be hoped will not be generally adopted. For, as well as this may apply to a few Asiatic dialects, it is only by violently wrenching words from their proper meaning, that it can be extended to the European members of the group. It is perhaps an even stronger objection to its use that some of the most distinguished philologists of the day have applied it specially to these Asiatic

* I have ventured in this word to omit the *o*, as is done in other words similarly formed from Greek and Latin; e.g. *magnanimous, philanthropy,* and *neuralgia,* which are never written *magno-animous, philo-anthropy,* und *neuro-algia.*

languages in contrasting them with the European.   Thus
L. Diefenbach, Or. Eur. p. 34 :

"Ihre beiden stämme in Asien : der Indische und der Iranische,
der wahrscheinlichst einst auch im östlichen Europa hauste, ritt,
und fuhr, bilden sämmtlichen Europäischen gegenüber eine
gruppe die wir die *Arische* nennen."

Other terms, such as "Japetic," "Indo-Germanic," and
"Caucasian," are too vague, or too limited.

But independently of the etymology of the names taken
by themselves, the question is ever arising, why they
should have been affixed to certain plants.   Where old
writers are quoted, and they give the reason for those
that they have themselves imposed, their authority is, of
course, conclusive ; but in other cases their notions are
often fanciful, and must be accepted with great reserve ;
for old as are the writers and their books, relatively to
modern botanists and floras, the names that they inter-
preted were often older than they, and the original meaning of them forgotten.   Synonyms in foreign languages,
including the Latin, are of essential service, but neither
are these very trustworthy ; for authors, mistaking the
sense of some unusual or obsolete word in one language,
have often translated it wrongly into another ; and this is
a fault that was as often made in ancient as in modern
times ; so that it is quite impossible to reconcile what is
said of certain plants by Greek and by Latin writers.   In
the case of the Hyacinth, Violet, Anemony, and other conspicuous flowers mentioned by Theocritus as Sicilian plants,

this is the more extraordinary, as the flora of that island is very similar to the Italian, and from its vicinity might have been familiar to Italian poets. But we find even in our own small island, that, what a Scotchman calls a "Bluebell," and makes the subject of popular songs, is a totally different flower from the English Bluebell.

It is this vague and random way of applying the same name to very different plants that occasions the greatest difficulty in the attempt to discover its original meaning. Who would dream that the Privet, for instance, has obtained a name indicative of "early spring" from having been confused under "Ligustrum" with the Primrose? or that the Primrose has borrowed its name from the Daisy? Numberless blunders of this kind arose while the art of describing a species was as yet unknown, and learned recluses, instead of studying nature in the fields, were perplexing themselves with a vain attempt to find in the north of Europe the Mediterranean plants of Theophrastus and Dioscorides. Indeed it was not till the publication of Turner's Herbal in the sixteenth century, that there was any possibility of ascertaining with certainty, through any English work, which of several species, or, indeed, which of several genera, might be meant by any given name; and, as it would be mere waste of time to attempt it now, the following vocabulary will contain, with the exception of a few from Chaucer, none but such as have been in use since that period.

Under the head of Popular Names our inquiry will

comprise those of the species most commonly cultivated in this country, as well as those of the naturalized and indigenous ones, but not Gardeners' or Farmers' names of mere varieties. Provincial words, that have not found their way into botanical works, are, with a very few exceptions, omitted. Many of these are very ancient, and expressive, and good names, and curiously illustrative of habits and superstitions that are rapidly passing away; but the study of them must be left to the local antiquary. They seem, generally, to be traceable to the language of the race which settled in the district where they prevail, and much less than the book names to a French or Latin source. In the northern counties and Scotland the nomenclature is very essentially different from that of the middle and south of England, and contains many words of Norse origin, and many of Frisian; but unfortunately these have been so vaguely applied, that nobody knows to what plants they, any of them, properly belong. This is more particularly the case with Scotch names. "Gowan," for instance, which in our English editions of Robert Burns is explained for us as "the daisy," means in different parts of Scotland many different plants, which agree in nothing but the having a yellow flower.* In Devonshire and the west of Somersetshire, there is also much that is peculiar, and, apparently, continued from the

---

* The Cleveland dialect of Yorkshire, a dialect almost purely Norse, has been most carefully investigated by the Rev. J. C. Atkinson in a work that is a model of accurate research, and should form a basis upon which to construct a more general glossary of the language of the North-Humbrian counties.

Anglo-Saxon period. In Suffolk, too, there has been a great number of valuable old names preserved, and carefully recorded in the Vocabularies of Moore and Forby. Many that are familiar to us in ancient herbals and in old poetry, have long fallen into disuse, except as they occur in the names of villages, and surnames of families, such as the places beginning with Gold, the ancient name of the marigold; as Goldby, Goldham, Goldthorpe, Goldsbury, and Goldworthy; and the families of Arnott, Sebright, Boughtflower, Weld, Pettigrew, Lyne, Spink, Kemp, and Harlock. Those of the commonest plants are the most variable, as the rarer ones have attracted too little of popular notice to have any but such as are given in books.

It seems desirable that these old names should be preserved, but there is already much greater difficulty in obtaining a correct list of those of any particular district, and the meaning of them, than there was a generation ago, from the dying out of the race of herb-doctors, and of the simplers, generally females, who used to collect for them. It is doubtful, indeed, whether any one of this class could now be found, who has learnt them from tradition, and independently of modern books.

One of the last was about eighty years ago living at Market Lavington in Wiltshire, a genuine old-fashioned specimen of his class, a Dr. Batter. He was understood to have had a regular medical education, probably as an apothecary, and certainly enjoyed a very high reputation. He has been described to me by a physician who knew him well, the late Dr. Sainsbury, sen., of Corsham, as a very unpretending man, and a successful practitioner, and visited and consulted from all parts of the county. He had been brought up very humbly, and lived and dressed as a poor man in a cottage by the road-side, where he was born, and where his father and grand-father had lived before him, and been famous in their day as bonesetters. There, if the weather permitted, he would bring out his chair and table, and seat his numerous patients on the hedgebank, and prescribe for them out of doors. It is said that, being well acquainted with every part of the county, he would usually add to the names of the plants that he ordered, the localities near the home of his visitor where they would most readily be found.

There were probably up to the end of the last century many such persons in other parts of England, combining the trades of herbarist and apothecary, and humbly supplying the place of those "gentlewomen" for whom Gerarde wrote his Herbal, and of the kind and charitable nuns of an earlier time. They were people of very humble or no education, and we might be tempted to suppose that we owe the absurd names we find in the following cata-logue to their ignorance and credulity. This is not at all the case. People in that rank of life seldom or never originate anything. Popular plant names, quite as much as popular tales, superstitions, ballads, and remedies, arise with a higher and more educated class of society, and

merely survive in a lower, after they have elsewhere
become obsolete. We can scarcely read without a smile of
scorn the meaning of such names as Fumitory, Devil's bit,
Consound, and Celandine; but it is to men of great
celebrity in their day, to Greek and Latin writers, such as
Theophrastus, Aristotle, Dioscorides, and Pliny, to Arabian
physicians, the most accomplished men of their time, and
to the authors and translators of our early herbals, that
we are indebted for nearly all such names as these. We
are not to criticize them, or attempt to explain them away,
but honestly to trace them back to their origin, and in
doing so to bear in mind, for our own humiliation, that
those who have betrayed such astonishing ignorance and
superstition, passed in their day for philosophers and men
of letters.

WORKS REFERRED TO.

Aasen, J.   Ordbog. o. d. Norske Folkesprog, 1850.
Adelung, J. C.   Wörterbuch, 1775.
Apuleius, L.   De herbarum virtutibus, Basil, 1528.
Atkinson, J. C.   The Cleveland Dialect, 1868.
Batman's Bartholomew de Glantvilla, 1582.
Bauhin, Casp.   Prodromus Theatri Botanici, 1620.
   „   De plantis a Sanctis nomen habentibus, 1591.
Bauhin, J.   Historia Plantarum, 1650.
Beckmann, J. A.   Lexicon Botanicum, 1801.
Bopp, F.   Comparative Grammar, 1862.
Bosworth, J.   Anglo-Saxon Dictionary, 1838.
Brunsfelsius, O.   Novum Herbarium, 1531.
Brunschwygk, H.   De arte distillandi, 1500.
Clusius, C.   Plantæ rariores, 1601.
Cockayne, O.   Leechdoms, 1864–6.
Coghan, Th.   Haven of Health, 1584.
Coles, W.   Adam in Eden, 1657.
   „   Art of Simpling, 1656.
Cordus, E.   In Dioscoridem, 1549.
Diefenbach, L.   Lexicon Comparativum, 1851.
   „   Origines Europææ, 1861.
Diez, Fr.   Etymologisches Wörterbuch, 1861.
Dodonæus, R.   Stirpium Historiæ, 1583.
Douglas, Gavin.   Virgil's Æneis, 1720.
Du Bartas, by Sylvester, Divine Weekes, 1611.
Du Cange, C.   Glossarium Mediæ Latinitatis, 1772–84.
Du Chesne, E. A.   Les Plantes Utiles, 1846.
Egger, E.   Dictionnaire Etymologique, 1870.
Evelyn, J.   Silva, 1786.
Fuchs, L.   Historia Stirpium, 1542.
Garnett, R.   Philological Essays, 1859.
Gerarde, J.   Herbal, 1597.
   „   ed. Th. Johnston, quoted as Ger. em., 1636.
Gesner, C.   De Lunariis herbis, 1668.

Glantvilla, Bar. by J. Trevisa, 1535; by Batman, 1582.
Graff, E. G. Althochdeutscher Sprachschatz, 1834.
Gray, S. F. Natural Arrangement, 1821.
Grete Herball, by Trevisa, 1526, and 1651.
Grimm, J. Gesch. d. Deutschen Sprache, 1848.
    ,,   J. and W. Deutsches Wörterbuch,
Halliwell, J. O. Archaic Dictionary, 1855.
Hampson, R. T. Medii ævi Kalendarium, 1841.
Herbarius,
Hill, J. Herbal, 1755.
Hogg, J. On the Classical Plants of Sicily, in Hooker's Journal, 1834.
Holmboe, C. A. Det Norske Sprog, 1852.
Honnorat, S. J. Dict. Provençal-Français, 1846.
Hyll, Th. Arte of Gardening, 1586.
Isidorus Hispalensis, de Etymologia, Migne's ed. 1850.
Jacob, E. Plantæ Favershamenses, 1777.
Jamieson, J. Scottish Dictionary, 1846.
Jennsen-Tusch, H. Folkelige Plantenavne, 1867.
Johnston, G. Botany of Eastern Border, 1853.
Keogh, W. Botanologia Hibernica, 1735.
Köne, J. R. Heliand, 1855.
Langham, W., Garden of Health, 1633.
Lightfoot, J. Flora Scotica, 1792.
Littré, E. Dict. de la langue Française, 1863– .
Lobel, M. Kruydtboek, 1581.
Lovell, R. Complete Herbal, 1665.
Lupton, Th. A Thousand Notable Things, 1595.
Lyte, H. Niewe Herbal, 1578.
Macer, Æm. De Virtutibus Herbarum, Basil, 1527 and 1581.
Matthioli, P. A. Comm. in Dioscoridem, Ven. 1554.
    ,,   Epitome aucta a Camerario, Frankf. 1586.
Mayer and Wright. National Antiquities, 1857.
Menzel, C. Index Nominum Plantarum, 1682.
Milne, Colin. Indigenous Botany, 1793.
Mone, F. J. Quellen und Forschungen, 1830.
Nares, R. Glossary, 1859.
Nemnich, P. A. Nomenclator multilinguis, 1793-98.
Newton, Th. Herbal to the Bible, 1585.
Ortus Sanitatis, by Cuba, 1486.
Outzen, N. Gloss. d. Friesischen Sprache, 1837.
Parkinson, J. Paradisus Terrestris, 1656.
    ,,   Theatrum Botanicum, 1640.
Plinius Secundus, Historia Naturalis, ed. Sillig, 1851.
Pott, A. F. Indogermanische Sprachen, 1853.
Promptorium Parvulorum, ed. Way, 1843–51.

Randolph, Frere.   Sloane MS. in Br. Mus. No. 3849, 1.
Ray, J.   Synopsis Stirpium, 1724.
Regimen Sanitatis Salernitanum, ed. A. Croke, 1830.
Ruellius, J.   De Natura Stirpium, 1724.
Scheler, A.   Dict. d'Etymologie Française, 1862.
Skinner, S.   Etymologicon, 1671.
Smith, W.   Dictionary of the Bible.
Stephanus, C.   De re hortensi, 1536.
Stockholm Medical MS. of 14th century in Archæologia, vol. xxx.
Tabernæmontanus, J. T.   Kraüterbuch, 1613.
Talbot, H. F.   English Etymologies, 1847.
Threlkeld, C.   Stirpes Hibernicæ, 1727.
Tournefort, J. P.   Institutiones Rei Herbariæ, 1719.
Tragus, H.   De Stirpibus, 1552.
Turner, R.   Botanologia, 1664.
Turner, W.   Herbal, 1551-1568.
        „   Names of Plants, 1548.
Ulfilas, ed. Massmann, 1857.
Wedgwood, H.   English Etymology, 1859-62.
Westmacott, W.   Scripture Herbal, 1694.
Winning, W. B.   Manual of Philology, 1838.
Wright, Th.   Domestic Manners in the Middle Ages, 1862.
        „   and Halliwell.   Reliquiæ Antiquæ, 1841-3.

ABBREVIATIONS OF LANGUAGES.

Ar.	Arabic.	M.Lat.	Middle-age Latin.
Arm.	Armenian.	Norw.	Norwegian.
A.S.	Anglo-Saxon.	O.E.	Old English of the 12th, 13th, and 14th centuries.
Boh.	Bohemian.		
Bret.	Breton.	O.Fr.	Old French.
Da.	Danish.	O.H.G.	Old High German.
Du.	Dutch.	Off.L.	Officinal Latin.
Er.	Erse or Old Irish.	O.N.	Old Norse, the ancient Danish.
Est.	Esthonian.		
Fin.	Finnic.	O.S.	Old Saxon of Lower Germany.
Fl.	Flemish.		
Fr.	French.	Pers.	Persian.
Fris.	Frisian.	Pol.	Polish.
Gael.	Gaelic of the Highlands.	Por.	Portuguese.
G.	German.	Rus.	Russian.
Go.	Gothic of Ulfilas.	Skr.	Sanskrit.
Gr.	Greek.	Scot.	Scotch of the Lowlands.
Heb.	Hebrew.	Slav.	Slavonian.
Ic.	Icelandic.	Sp.	Spanish.
It.	Italian.	Sw.	Swedish.
L. or Lat.	Latin.	Tar.	Tartar.
Lap.	Lappish.	Wal.	Walachian.
Lith.	Lithuanian.	W. or Wel.	Welsh.
L.Ger.	Low German, Plattdeutsch.	Zend.	The old language of Persia, probably the Mede dialect.
M.Gr.	Modern Greek.		

NAMES OF BRITISH PLANTS.

AARON, a corruption of L. *arum*, Gr. ἀρον, into a more familiar word, A. maculatum, L.

ABELE, Du. *abeel*, in Pr. Pm. *awbel* or *ebelle*, from Fr. *aubel*, M.Lat. *albellus*, whitish, a word that occurs as the name of the tree in Lambertus Ardensis, p. 79 : " *Albellus* cum tilia juxta crucem, ubi plantata est ad peregrinatorum requiem et præsidium," and which refers to the white colour of the twigs and leaves. Our *Abele* is this Dutch name, *abeel*, with which it was introduced from Holland in Evelyn's time (Silva, 1, 207). Populus alba, L.

ACACIA, Gr. ἀκακια, guilelessness, good nature ; a name given by Dioscorides (b. i. ch. 130) to a small Egyptian tree, but now transferred in popular language to an American Robinia, R. Pseudacacia, L.

ACH, Fr. *ache*, the old name of parsley, from L. *apium*, formed by the change of *pi* to *ch*, as in *sapiam* to *sache*, *propius* to *proche*, etc., now only retained in *Smallage*, the small ach, Fr. *ache de marais, ache rustique, ache femelle*, as contrasted with the Alexander, Fr. *ache large, grande ache*, Apium, *L.*

ACONITE, derived by Theophrastus from the village 'Ακοναι, but by Ovid (Met. vii. 419) from growing upon rock, ἀκονη,

" Quæ quia nascuntur dura vivacia caute,
  Agrestes *aconita* vocant."

Pliny suggests that it is so called from growing where there is no dust, *ἀ*, not, and *κονις*, dust, " nullo juxta ne pulvere quidem nutriente." It is, rather, a word of the same derivation, but used in a different sense, *ἀκονιτον*, without a struggle, alluding to the deadly virulence of its juice, which W. Turner says "is of all poysones the most hastie poysone." The plant of the Greek writers has been identified with the monkshood,

Aconitum Napellus, L.

,, WINTER-, Eranthis hyemalis, D.C.

ADDER'S TONGUE, from the Du. *adderstong*, in old MSS. called *nedderis gres* (grass) and *nedderis-tonge*, M. Lat. *serpentaria*, from its spike of capsules having some fancied resemblance to that reptile's tongue,

Ophioglossum vulgatum, L.

ADDERWORT, the snakeweed or bistort, from its writhed roots, Polygonum Bistorta, L.

AFFADYL, M.Lat. *affodillus*, from L. *asphodelus*, Gr. *ἀσφοδελος*, pictured in Ort. Sanit. as an Iris, an old term replaced in later times by *Daffodil*,

Narcissus Pseudonarcissus, L.

AGARICK, L. *agaricum*, G. *ἀγαρικόν*.

AGRIMONY or EGREMONY, in Chaucer *egremoine*, L. *agrimonia*, a word of uncertain origin, probably the Gr. *ἀργεμωνη*; but what this plant was, or why it was called so, is unknown. A. Eupatorium, L.

,, HEMP-, from its being confused under the name of *Eupatoria* with the preceding species, and the resemblance of its leaves to those of hemp, Eupatoria cannabina, L.

ALBESPYNE, in old works the name of the white- or haw-thorn, from the whiteness of its rind as contrasted with that of the black-thorn, Fr. *aubespine*, M.Lat. *albepinus*, from L. *alba*, white, and *spina*, thorn,

Cratægus Oxyacantha, L.

ALDER, formerly, and still locally, and more properly,

ALLER, A.S., *ælr, alr, aler,* with a *d* inserted for euphony, as in *alderfirst, alderlast,* Go. *erila,* whence G. *erle,* O.H.G. *elira,* L. *alnus.* Garnett would connect it with words implying moisture, as *uligo, ulva,* etc. (Phil. Ess. p. 30). The similarity of the Danish *elle* with the name of fairies in that language, *elle-trä,* and *elle-folk,* has misled Goethe to give the name of *erlen-könig* to the fairy-king. There is no etymological connection between the two.

Alnus glutinosa, L.

„ BLACK-, or BERRY-BEARING-, a buckthorn that was wrongly associated by the older botanists with the alder, but distinguished from it by bearing berries,

Rhamnus Frangula, L.

ALECOST, from L. *costus,* some unknown aromatic, and *ale,* so called from its having formerly been esteemed an agreeable aromatic bitter, and much cultivated in this country for flavouring ale. "Put certaine handfuls of this herbe in the bottom of a vessel, and tunne up new ale upon it." Coghan (ch. 71.)—See COSTMARY.

Balsamita vulgaris, L.

ALE-HOOF, ground-ivy, from *ale,* and *hoof,* which appears to be the A.S. *hufe,* crown, Du. *huif,* O.N. *hufa,* and to have been given to this plant either as translating its Gr. and Lat. names στεφανωμα γης, *corona terræ;* or in allusion to the chaplet that crowned the alestake at a public-house, as in Chaucer (Prol. l. 667):

> "A gerlond hadde he sette upon his hede,
> As gret as it were for an ale-stake."

J. and W. Grimm would regard it as a compound of *ei,* ivy, and *loof,* leaf, but, as we shall see under IVY, the bush has been so named from the herb, and not the herb from the bush.—See GILL. Nepeta Glechoma, Benth.

ALEXANDERS or ALISANDERS, the horse-parsley, from its Latin specific name *Alexandrinum,* a name that Ray says was given to it from its being a plant of *Alexandria* in

Egypt, but more probably derived from an earlier name, *Petroselinum Macedonicum*, a parsley of Macedon, Alexander's country.—See Dodoens, p. 697.

Smyrnium Olus atrum, L.

ALISSON or ALISON, L. *Alyssum*, from Gr. *ἀ*, not, and λυσση, madness, a plant called so by the ancients, because, as Turner says, "it helpeth the biting of a wod dogge." He seems, by his description of it, to mean the field madder, Sherardia arvensis, L. a very different plant from that which now bears the name. That of modern botanists is a genus of Cruciferæ, Alyssum, L.

„ SWEET-, from its scent of honey,

A. maritimum, L.

ALKANET, Fr. *orcanette*, a dim. formed from It. and Sp. *alcana*, representing the Arabic *El* or *Al hanne* or *canne*, the name of a very different plant, the Lawsonia inermis, L. which yields, like the alkanet, a red dye, the *Henna* of the harems, Anchusa officinalis and tinctoria, L.

„ BASTARD-, Lithospermum arvense, L.

ALKEKENGI, from the Arabic, Physalis Alkekengi, L.

ALL-BONE, a name taken from the Gr. ὁλοστεον, as a compound of ὁλος, whole, and ὁστεον, bone, applied to a very tender plant, "whereof," says Gerarde, "I see no reason, unlesse it be by the figure called Antonomia ; as, when we say in English, ' He is an honest man,' our meaning is, ' He is a knave.' " The ὁλοστεον of Dioscorides was some other, probably, very different plant, the name of which has been transferred to this from its jointed skeleton-like stalks. In Cheshire it is called " Breakbones" from their snapping off at the joints.

Stellaria Holostea, L.

ALLELUJAH, see HALLELUJAH.

ALLGOOD, Du. *algoede*, G. *allgut*, from a Latin name *tota bona*, Fr. in Cotgrave and Palsgrave *toutte bonne*, given in old works to a goosefoot, that is otherwise called " English Mercury," on account of its excellent qualities as a remedy and as an esculent; whence the proverb:

"Be thou sick or whole, Put Mercury in thy koole."

Coghan (ch. 29).

Chenopodium Bonus Henricus, L.

ALL-HEAL, see CLOWN'S ALL-HEAL.

ALL-SEED, from the great quantity of its seed,

Radiola Millegrana, Sm.

and also Chenopodium polyspermum, L.

and Polycarpon tetraphyllum, L.

ALSIKE, Sw. *alsike-klöver*, a clover so called from its growing abundantly in the parish of Alsike near Upsal in Sweden (J. H. Lunden in N. and Q.)

Trifolium hybridum, L.

AMADOU, from the Fr., a word of uncertain derivation,

Polyporus igniarius, Fries.

AMARANTH, Gr. ἀμαραντος, from ἀ not, μαραινω, wither, a word of not unfrequent occurrence in Milton and other poets, as a vague name for some unfading flower. The original species was one that, from its quality of reviving its shape and colour when wetted with water, was much used by the ancients for winter chaplets. The phrase in St. Peter's 1st Epis. ch. v. 4, " a crown of glory that fadeth not away," is in the original, "the amarantine crown of glory," τον ἀμαραντινον της δοξας στεφανον. The plants which botanists call so are the species of the genus to which the "Love lies bleeding" belongs. Amarantus, L.

AMBROSE, a name given in old writings to some sweet-scented herb, from Gr. ἀμβροσια, the food of immortals, Skr. *amrita*, elixir of immortality, from *a*, not, and *mri*, Lat. *mori*, die. It is uncertain what plant was meant by the Greek term, but whatever this may have been, Matthioli tells us in his Comment on Dioscorides, (l. iii. c. 12,) that it was called so by the ancients, because a continued use of it rendered men long-lived, in the

same way as the *Ambrosia,* which was the food of the
Gods, was fabled to preserve them immortal. The Pr. Pm.
translates it *salgia sylvestris,* wild sage ; Palsgrave *ache
champestre,* field parsley ; Cotgrave *oke-of-Cappadocia,* or
-of-Jerusalem. The name is now assigned by botanists to
a plant of the wormwood kind, Ambrosia, L.
That of the poets was discovered by Isis :

> Ἑυρειν δε ἀυτην και το της ἀθανασιας φαρμακον.—Diod. Sic. i. 25.

In Homer and the other early poets ambrosia was the food,
and nectar the drink, of the Gods, as in Odyss. v. 93 :

> Ὡς ἀρα φωνησασα θεα παρεθηκε τραπεζαν,
> Ἀμβροσιης πλησασα· κερασσε δε νεκταρ ἐρυθρον.

and in Ovid (ex Ponto, i. 10)

> "Nectar et ambrosiam, latices epulasque deorum."

But the two became confused together, and the same author
tells us (Met. xiv. 606) that Venus, after bathing Æneas,

> "Ambrosia cum dulci nectare mista,
> Contigit os, fecitque deum."

In the beautiful tale of Cupid and Psyche in the Golden
Ass of Apuleius, (b. vi,) Jupiter in conferring immortality
on Psyche, gives her a cup of ambrosia to drink : "por-
recto *ambrosiæ* poculo, Sume, inquit, Psyche, et immortalis
esto." The *Ambrose* of our older English writers seems to
have been Chenopodium Botrys, L.

Ameos, the genitive of *Ammi,* used, like *Caruy,* for the
seed of the plant, A. majus, L.

Anemony, Gr. ἀνεμωνη, from ἀνεμος, Skr. *anila,* wind,
from *an,* to blow. It is said by Bion to have sprung from
the tears that Venus wept over the body of Adonis, a
myth that seems to whisper that the tears of that frail and
loving goddess were soon blown away.—(Idyl. i. l. 62).

> Ἀι ἀι ταν Κυθερειαν, ἀπωλετο καλος Ἀδωνις.
> Δακρυον ἀ Παφια τοσσον χεει, ὁσσον Ἀδωνις
> Αἱμα χεει· τα δε παντα ποτι χθονι γιγνεται ἀνθη.
> Αἱμα ρςδον τικτει, τα δε δακρυα ταν ἀνεμωναν.

Alas the Paphian! fair Adonis slain!
Tears plenteous as his blood she pours amain.
But gentle flowers are born, and bloom around,
From every drop that falls upon the ground:
Where streams his blood, there blushing springs the Rose,
And where a tear has dropp'd a Wind-flower blows.

Whether the flower that we now call *Anemony*, was that which the Sicilian writers meant, is a question, into which it were here out of place to enter. Pliny tells us (H.N. l. xxi. c. 11) that it was so named, because it never opens but when the wind is blowing. Ovid describes it as a very fugacious flower, and after comparing it with that of the Pomegranate, says (Met. x. 737):

"Brevis est tamen usus in illis,
 Namque male hærentem, et nimia levitate caducum
 Excutiunt idem qui præstant nomina venti."

It is doubtful whether he meant the same plant as Pliny; and he could scarcely have meant that which we call so now; more probably a cistus, or rock-rose. The name is now applied to the genus   Anemone, L.

ANET, dill seed, from L. *anethum*, Gr. ἀνηθον,

                        A. graveolens, L.

ANGELICA, its Lat. name, either as Fuchs tells us, (Hist. Plant. p. 126,) "a suavissimo ejus radicis odore, quem spirat," or "ab immensa contra venena facultate," from the sweet scent of its root, or its value as a remedy against poisons and the plague, yielding, as Brunschwygk tells us, "das aller-edelst wasser das man haben mag für die pestilenz;" and of which Du Bartas says, (Third day, p. 27,) Sylvester's translation, 1641,

"Contagious aire ingendring Pestilence
 Infects not those that in their mouths have ta'en
 Angelica, that happy counterbane
 Sent down from heav'n by some celestial scout,
 As well the name and nature both avowt."

                        Angelica sylvestris, L.

ANISE, or as in "The Englishman's Docter," ANNY,

"Some *Anny* seeds be sweet, and some more bitter."

L. *anisum*, Gr. ἄνισον or ἄνησον; the *Anny* having arisen from a mistake of *Anise* for a plural noun ;

Pimpinella Anisum, L.

ANTHONY, ST. his nut and turnep, see under SAINT A.

APPLE, A.S. *æpl, æppel*, O.N. *epli*, Sw. *æple*, Da. *æble*, G. *apfel*, O.H.G. *aphol*, Wel. *afal*, derived from a more ancient form, *apalis*, preserved in the Lith. *obolys*, or *obelis*. Lett. *ahboli*. In all the Celtic and Sclavonian languages the word is, with allowance for dialect, the same. This similarity, or, we may say, identity of name, among alien nations would lead us to believe that it was brought with the tree from some one country, and that, no doubt, an Eastern one; and that the garden apple is not, as it is often supposed to be, merely an improved crab, but rather the crab a degenerate apple. This was, apparently, the only fruit with which our ancestors were acquainted, before they came into Europe; for, with the exception of a few wild berries and the hazel nut, it is the only one for which we have a name that is not derived from the Latin or French. It seems to have accompanied them on a northern route from the western spur of the Himalayan mountains, a district extending through Ancient Bactria, Northern Persia, and Asia Minor, to the Caucasus, and one from which we have obtained, through the Mediterranean countries, and within the historical period, the peach, apricot, plum, damson, cherry, filbert, vine, and walnut, and probably some of the cereal grains ; a district in which there is reason to think that our portion of the human race first attained to civilization, and whence it spread, with its domestic animals and plants, to the south-east and north-west. The meaning of the word is unknown. It is very possibly from Skr. *amb*, eat, and *p'hal*, fruit, but as *ap* is, in Zend and Sanskrit, "water," we might be tempted to believe that it originally meant "water-fruit," or "juice-fruit," with which the Latin *pomum*, from *po*, to drink, exactly tallies. The remarkable coincidences of name, to which allusion has been made, are due to the intimate connection with each other of all the Ind-European nations and their languages, from their having grown up in the same nursery together in Upper Asia, and dispersed subsequently to their becoming acquainted with this fruit; and not to a mutual borrowing of it since their settlement in Europe. Pyrus Malus, L.

APRICOT, in Shakspeare (M.N.D. iii. 1) APRICOCK, in older writers, ABRICOT and ABRECOCKE, It. *albericocca* and *albicocco*, from Sp. *albaricoque*, Ar. *al burqûq* or *barkokon*, from Mod. Gr. βρεκκοκα, O. Gr. of Dioscorides and Galen, πραικοκκια, L. *præcoqua* or *præcocia*, early, from the fruit having been considered to be an early peach. A passage from Pliny (Hist. Nat. xv. 12) explains its name. "Post autumnum maturescunt Persica, æstate *præcocia*, intra xxx annos reperta." There is a good paper upon it in Notes and Queries, Nov. 23, 1850. "The progress of this word," says the author, "from W. to E., and then from E. to S.W., and thence to N., and its various changes in that progress, are strange. One would have supposed that the Arabs living near the region of which the fruit is a native, might have either had a name of their own for it, or at least have borrowed one from Armenia. But they have apparently adopted a slight variation of the Latin. The Spaniards must have had the fruit in Martial's time, [who alludes to it in the words :

'Vilia maternis fueramus præcoqua ramis,
Nunc in adoptivis persica cara sumus.'
—Lib. xiii. Ep. 46.]

but they do not take the name immediately from the Latin, but through the Arabic, and call it *albaricoque*. The Italians again copy the Spanish, not the Latin, and call it *albicocco*. The French from them have *abricot*. The English, though they take their word from the French,

at first called it *abricock* (restoring the *k*), and lastly with
the French termination, *apricot*." Prunus armeniaca, L.

ARACH, in Pr. Pm. and in Palsgrave ARAGE, the older
spelling of ORACH.

ARCHAL, a lichen called more commonly *Orchil*,
Roccella tinctoria, D.C.

ARCHANGEL, M.Lat. *archangelica*, so called Parkinson
tells us, "ab eximiis ejus viribus;" Nemnich, from its
having been revealed by an angel in a dream ; more probably from its being in blossom on the Archangel St.
Michael's day, the 8th of May, old style, and thence supposed to be a preservative against evil spirits and witch-
craft, and particularly against the disease in cattle called
elfshot, G. *hexenschuss*, ulcera regia. The name is applied
to an umbelliferous plant,   Angelica archangelica, L.
and to certain labiates, severally called

,,	RED-,	Stachys sylvatica, L.
,,	WHITE-,	Lamium album, L.
,,	YELLOW-,	Lamium Galeobdolon, Cr.

ARNUT, or ERNUT, Du. *aardnoot*, the earth-nut,
Carum Bulbocastanum, K. and Bunium flexuosum, With.

ARROW-GRASS, a translation of its Greek name, *triglochin*,
from the three points of its capsules, τρεις, three, γλωχις,
arrow-point,   T. palustre, L.

ARROW-HEAD, from the shape of the leaves,
Sagittaria sagittifolia, L.

ARSMART, Fr. *curage*, i.e. *cul-rage*, the water-pepper,
from the irritating effect of the leaves,
Polygonum Hydropiper, L.

ARTICHOKE, in Turner ARCHICHOCKE, Fr. *artichaux*, It.
articiocco, Sp. *artichofa*, a name which Diez derives from
Ar. *ardischauki*, earth-thorn, and which was introduced
with the plant by the Moors of Spain.
Cynara Scolymus, L.

ASARABACCA, a name adopted, as a compromise or middle

term, in consequence of the confusion between the two plants, *Asarum* and *Baccharis,* one with the other. "In former times," says Parkinson (Th. Bot. p. 115), "divers did thinke that Asarum and Baccharis in Dioscorides were all one hearbe, and thereupon came the name of Asarabaccara; some taking Asarum to be Baccharis, and so contrarily some taking Baccharis to be Asarum."

Inula Conyza, D.C. and Asarum europæum, L.

ASH, A.S. *æsc,* Da. and Sw. *ask,* O.N. *askr,* O.H.G. *asc,* G. and Du. *esche.* From the toughness of the wood it was much used for spear-shafts, and A.S. *æsc* came to mean a spear, and *æsc-plega,* the game of spears, a battle. *Fresne* in the same manner was used in France for a spear, whence the expression *brandir le fresne.* It was further extended to the man who bore it, and he was himself called *æsc.* Being also the wood of which boats were built, the A.S. *æsc* and O.N. *askr* meant a vessel, just as a barge with an oak bottom is called, from its wood, in L.Germ. *eeke,* Du. *æke,* Sw. *eka.* The derivation and primary meaning of *Ash* is obscure. It is not improbably connected with L. *ascia,* Gr. ἀξινη, and *axe,* and with L. *axis,* an axle, from the tough wood of this tree having in all times been preferred for axe handles and axletrees.

Fraxinus excelsior, L.

„ MOUNTAIN-, the rowan, from a fancied resemblance of its pinnate leaves to those of the ash-tree, and its usual place of growth, Pyrus aucuparia, Gärt.

ASH-WEED, AISE- or AX-WEED, from its ternate leaves somewhat resembling those of the *ache* or celery. See ACHE. Ægopodium Podagraria, L.

ASPARAGUS, Gr. ἀσπαραγος, A. officinalis, L.

ASPEN, the adjectival form of ASPE, the older and more correct name of the tree, and that which is used by Chaucer and other early writers; A.S. *æpse* and *æsp,* G. *aspe,* O.H.G. *aspa,* O.N. *espi,* Populus tremula, L.

ASPHODEL, Gr. ἀσφοδελος, a word of unknown derivation, applied in Homer (Odyss. xi. 539) as an epithet to a meadow, ἐν ἀσφοδελῳ λειμωνι. The plant so called by Greek writers of a later age, was one that had edible roots, that were laid in tombs to be food for the dead, and is that to which Charon alludes in Lucian's Καταπλους, c. 2 : " I know," says he, " why Mercury keeps us waiting so long. Down here with us there is nothing to be had but asphodel, and libations, and oblations ; and that in the midst of mist and darkness ; but up in heaven it is all bright and clear, and plenty of ambrosia there, and nectar without stint." This root, under the name of *cibo regio*, food for a king, was highly esteemed in the middle ages, but however improved by cultivation, it is likely to have been troublesome by its diuretic qualities, and has probably for that reason gone out of fashion. There is some ground to suspect that it was the original claimant of an expressive name that has since passed to the dandelion.

The plant of the Greek poets is supposed to be the
Narcissus poeticus, L.
That of Lucian and of modern botanists, Asphodelus, L.
That of our earlier English and French poets,
Narcissus Pseudonarcissus, L.
  ,,   Bog-, or LANCASHIRE-,
Narthecium ossifragum, Huds.
  ,,   SCOTCH-,   Tofieldia palustris, Huds.

ASSES-FOOT, Fr. *pas d'âne*, the colt's foot, from the shape of the leaf,   Tussilago Farfara, L.

ASS-PARSLEY, in old works given as the translation of Fr. *cicutaire*, the same, probably, as fools-parsley,
Æthusa Cynapium, L.

ASTER, Gr. ἀστηρ, a star, from the radiate flower,
Aster, L.

AUTUMN-BELLS, from its bell-shaped flowers and their season of opening,   Gentiana Pneumonanthe, L.

AVENS, in Pr. Pm. *avence*, in Topsell and Askham *avance*, M.Lat. *avantia* or *avencia*, in Ort. San. *anancia*, a word of obscure origin, and quite unintelligible, spelt also *auartia*, *anantia*, *arancia*, and *amancia*,

,,	COMMON-, or YELLOW-,	Geum urbanum, L.
,,	MOUNTAIN-,	Dryas octopetala, L.
,,	WATER-, or NODDING-,	Geum rivale, L.

AVEROYNE, of the Stockholm Med. M.S., but long disused, Fr. *aurone*, from Lat. *abrotanum* (Scheler), the southernwood, Artemisia Abrotanum, L.

AWL-WORT, from its subulate leaves,

Subularia aquatica, L.

AYE-GREEN, ever-green, a translation of Lat. *sempervivum*. *Aye* is the A.S. *æg*, ever, properly an egg, which, having no beginning or end, was symbolical of eternity, Go. *aiv*, L. *æ* in *ævum*, *ætas*, and *æternus*, Gr. *ἀεί*. The plant so called from its conspicuous tufts of evergreen leaves, the houseleek, is Sempervivum tectorum, L.

BACHELOR'S BUTTONS, a name given to several flowers "from their similitude to the jagged cloathe buttons, antiently worne in this kingdom," according to Johnson's Gerarde, p. 472, but ascribed by other writers to "a habit of country fellows to carry them in their pockets to divine their success with their sweethearts;"
usually understood to be a double variety of Ranunculus, according to others, of Lychnis sylvestris, L.
in some counties, Scabiosa succisa, L. .

BALDMONEY, or BAWD-MONEY, the mew, a corruption of L. *valde bona*, very good. The Grete Herball, ch. ccccxxxiii, speaking of Sistra, says, "Sistra is dyll, some call it Mew, but that is not so. Howbeit they be very like in properties and vertue, and be put eche for other, but Sistra is of more vertue than Mew, and the leaves be lyke an herbe called *valde bona,* and beareth small sprigges as spiknarde. It groweth on hye hylles."

Meum athamanticum, L.

in some authors, incorrectly,   Gentiana lutea, L.

BALLOCK-GRASS, A. S. *bealloc-wyrt,* from its tubers resembling small balls, whence its Greek name, ὄρχις,

Orchis, L.

BALM, BAULM, or BAWM, contracted from *Balsam,* L. *balsamum,* by some said to be derived from Hebr. *bol smin,* chief of oils, by W. Smith from Hebr. *bâsâm,* balm, and *besem,* a sweet smell, terms originally applied to a plant very different from that which now bears the name,

Melissa officinalis, L.

„ BASTARD-,   Melittis Melissophyllum, L.

BALSAM, or BALSAMINE, see above,

Impatiens Noli me tangere, L.

BANEBERRY, A.S. *bana,* murderous, from its poisonous quality. Hill says in his Herbal (p. 320), that children who have eaten the fruit have died in convulsions.

Actœa spicata, L.

BANEWORT, from its being supposed, like several other marsh plants, to bane sheep, and Salmon tells us that it does so by ulcerating their entrails.

Ranunculus Flammula, L.

BANKCRESS, from its growing in hedge banks, the hedge mustard,   Sisymbrium officinale, L.

BARBARA, ST. her cress, see under ST. B.

BARBERRY or BERBERRY, M.Lat. *berberis,* Ar. *barbāris,*

B. vulgaris, L.

BARLEY, called in Sloane MS. No. 1571, 3, at fol. 113, *barlych,* and in the A.S. Chronicle, A.D. 1124, *bærlic,* from *bær,* which represents both the A.S. *bere,* barley-corn and *beor,* the liquor brewed from it, and *lic* for *leac,* plant, a name identical in meaning with the *bær-cræs* of Ælfric's vocabulary. Verstegan says that the name of *barley* was given to it by reason of the drinke therewith made called *beere*, and from *beerlegh* it came to *berlegh*, and from *berlegh* to *barley*. It would seem that, as the language became corrupted by the settlement of Danes and French in the country, and the vowels less correctly pronounced, the *lic* was added to prevent confusion. The dictionary derivation of it from the Welsh *barlys* is untenable, both for philological reasons, and for that it is highly improbable that the English of the twelfth century would have borrowed from a half-civilized mountain race a name for a familiar plant. See BEAR. Hordeum vulgare, L.

,, WALL-, see MOUSE BARLEY.

BARNABY-THISTLE, from its flowering about the time of St. Barnabas' day, the 11th of June, old style, which corresponds to the 22nd June of the new.

Centaurea solstitialis, L.

BARREN-WORT, called so, says Gerarde, p. 389, "because it is an enemy to conception, and not because it is described by Dioscorides as being barren both of flowers and leaves." Nevertheless, this belief in its sterilizing powers may be due to the remark of Dioscorides, who must have meant some other plant, for this seeds very freely in Styria and other parts of Austria.

Epimedium alpinum, L.

BASE-BROOM, L. *Genista humilis*, a name that does not, as its Latin synonym would lead us to suppose, refer to its lowly growth as compared with that of the common broom, but to its being used as a *base* to prepare woollen cloths for the reception of scarlet and other dyes.

Genista tinctoria, L.

BASE-ROCKET, a mignonette so called from its rocket-like leaves, and its being used as a *base* in dyeing woollen cloths. See Aubrey's Wilts (ed. Jackson), p. 50.

Reseda lutea, L.

BASIL, Gr. βασιλικον, royal, probably from its being used in some royal unguent, bath, or medicine,

Ocymum basilicum, L.

„ FIELD-,   Calamintha Clinopodium, Bent.

BASIL-THYME, so called, says Parkinson (Th. Bot. p. 19), "because the smell thereof is so excellent, that it is fit for a king's house;"   Calamintha Acinos, Clair.

BASSINETS, Fr. *bassinet*, a small basin, a skull-cap, from the shape of the flower,   Ranunculus, L.

„ BRAVE-, in Lyte's Herball, Caltha palustris, L.

BAST-TREE, the lime-tree, from its inner bark furnishing bast for matting, a word introduced with the material from Germany or Denmark, and related to Skr. *pas*, bind, Pers. *benden*, and Zend and Skr. *bandh*, whence Z. *basta*, bound.

Tilia europæa, L.

BAY, Fr. *baie*, formed, by the usual omission of *c* between two vowels, from L. *baca*, often so spelt for *bacca*, a berry. In old works *bay* means a berry generally, as "the *bayes* of ivyne," but as those of the sweet bay, the *lauri baccas* of Virgil, were an article of commerce, the term came to be applied to them exclusively, and was thence extended to other evergreens, much as is *laurel* at the present day.

Laurus nobilis, L.

„ DWARF-,   Daphne laureola, L.

BEAM-TREE, more properly WHITE BEAM, from A.S. *beam*, a tree, O.S. *bam*, G. *baum*, Goth. *bagms*, O.N. *ba͠mr*, words related to G. *bauen*, build. *Beam*, without the *White* prefixed, is a vague term, meaning in A.S. a tree generally, so that *Beam-tree* is a silly pleonasm, a *tree-tree*.   Pyrus Aria, L.

BEAN, A.S. *bean*, O.N. *baun*, Du. *boon*, G. *bohne*, Da. *bönne*, Sw. *böna*, is considered by W. and J. Grimm to be connected with Lat. *faba* ; and there are in different languages words in some degree intermediate between the two ; not, however, that the Germans derived their word from the Latin. The plant probably came from an Eastern country, and by the northern, and not the Mediterranean route, and the name with it. Vicia Faba, L.

BEAR, BEER, or BERE, A.S. *bær*, Fris. *bar*, barley, a grain that might seem to have been so called from *beer*, the liquor brewed from it, a word for which L. Diefenbach remarks that " sichere Etymologien fehlen noch." Outzen and several other philologists support this derivation of it, but J. Grimm would trace it to Go. *bairan*, bear, whence *baris*, gen. *barizis*, O.N. *barr*, gen. *bars*. Gesch. d. Deutsch. Sprache, i. p. 65. It may be related to Hind. *bagra*, a kind of millet (holcus spicatus, L.), that is much cultivated in the mountains of the north of India. In our northern counties *bear* means the four-rowed variety of barley and *bigg* the six-rowed. Brockett, p. 25. Hordeum vulgare, L.

BEARBERRY or BEAR'S BILBERRY, from its fruit being a favourite food of bears, Arctostaphylos uva ursi, L.

BEAR-BIND or BARE-BIND, from binding together the stalks of bear or barley, Convolvulus arvensis, L.

BEAR'S-BREECH, from its roughness, a name transferred by some mistake from the acanthus to the cow-parsnip, Heracleum Sphondylium, L.

BEAR'S-EARS, from its former Latin name, *ursi auricula*, in allusion to the shape of its leaf, Primula Auricula, L.

BEAR'S-FOOT, from its digitate leaf, Helleborus fœtidus, L.

BEAR'S-GARLICK, so called, says Tabernæmontanus, "quia ursi eo delectantur," Allium ursinum, L.

BEARWORT, from the G. *bärwurz*, which Adelung suggests is rather to be derived from its use in uterine complaints than from the animal, Meum athamanticum, L.

BEDE-SEDGE, from its round bead-like burs, resembling the beads used by Roman Catholics and Buddhists for counting their prayers, A.S. *bead*, a prayer, a name given to it by Turner, Sparganium ramosum, L.

2

BEDSTRAW, or OUR LADY'S BEDSTRAW, L. *Stæ. Mariæ stramen*, G. *unser lieben frauen bettstro*, from its soft, puffy, flocculent stems and golden flowers, a name that refers to straw having formerly been used for bedding, even by ladies of rank, whence the expression of their being "in the straw." Thus in the Latin hymn, No. 128 in Bäsler's Altchristliche Lieder, Mary sings :

"Dormi nate bellule, Stravi lectum *foeno* molli : Dormi mi animule."

The name may allude more particularly to her having given birth to her son in a stable, with nothing but wild flowers for her bedding.   Galium, L.

„   WHITE-,   G. Mollugo, L.

„   YELLOW-,   G. verum, L.

In old writers its Latin and German synonyms are given equally to the wild thyme. See C. Bauhin de plantis sanctis, p. 71.

BEE-LARKSPUR, "from the resemblance of its petals studded with yellow hairs to a bumble bee whose head is buried in the recesses of the flower." Treas. of Bot.

     Delphinium grandiflorum, L.

„   -NETTLE, from its nettle-like leaves, and the supposed fondness of bees for its flowers,

     Galeopsis Tetrahit, L.

„   -ORCHIS or BEE-FLOWER, from the resemblance of its flower to a bee,   Ophrys apifera, L.

BEE'S-NEST, from the nest-like compact growth of its inflorescence,   Daucus Carota, L.

BEECH, A.S. *boc, bece, beoce*, Go. *boka*, M.H.G. *buoche*, O.H.G. *puocha*, G. *buch*, Du. *beuk*, O.N. *beyki*, Da. *bög*, Sw. *bok*, words which, in their several dialects, mean, with difference of gender only, a book and a beech-tree, from Runic tablets, the books of our ancestors, having been made of this wood.   Fagus sylvatica, L.

BEET, L. *beta*, from the resemblance of the seed to the

second letter of the Greek alphabet. There are some verses of Columella to this effect, which are quoted by Fuchs and Parkinson :

> " Nomine cum Grajo, ceu litera proxima primæ
> Pangitur in cera docti mucrone magistri,
> Sic et humo pingui ferratæ cuspidis ictu
> Deprimitur folio viridis, pede candida beta."

Nemnich pronounces this idea to be a mere *grille* or fancy, but gives no better derivation for the word.

B. maritima, L.

BEGGAR'S LICE, from its burs sticking to the clothes, and somewhat resembling those vermin,

Galium Aparine, L.

BELL-FLOWER, from the shape of the corolla,

Campanula, L.

BELLADONNA, It. *bella-donna*, fair lady, the deadly night-shade, called so, according to Tournefort, and to G. Burnet (Outl. of Botany, 4514), and to Duchesne (Pl. ut. p. 90), from its berries, known in France as *guines de côtes*, being used by the Italian ladies as a cosmetic. Ray also says that it was called *belladonna*, " quia ex ejus succo sive aqua destillata fucum conficiunt fœminæ, quo faciem oblinunt, et ex rubicunda pallidam efficiunt frigoris vehementia." (Cat. Plant. Cant. p. 43.) Atropa Belladonna, L.

BELLEISLE-CRESS, why called so, unknown,

Barbarea præcox, R.B.

BEN or WHITE BEN, from a supposed Ar. *behen*, which however is not to be found. Silene inflata, L.

BENNET, see HERB B.

BENT-GRASS, any wiry grass, such as usually grows upon a *bent*, *i.e.* a common, or other neglected broken ground, a word often used in that sense in old English poetry—as : " Bowmen bickered upon the *bent*." Chevy Chase,—and preserved in Scotland to this day. Jamieson refers it to G. *binse*, a rush, but the similarity of these words seems to be an accidental coincidence, and the name of the grass to have been taken from its place of growth, as in the case of *heath, brake,* and *brier,* and not the ground called so from the grass.   Under the name of *Bent* are comprised Agrostis vulgaris, L.   Triticum junceum, L.   Arundo arenaria, L. and many more.

BERGAMOT, a name popularly given to a common garden plant,   Monarda fistulosa, L.

BERTRAM, a corruption of L. *pyrethrum,* that seems to have been adopted from German writers,

P. Parthenium, L.

BERRY ALDER, a buckthorn that was once wrongly associated with the alders, and distinguished from them by bearing berries,   Rhamnus Frangula, L.

BETONY or WOOD BETONY, L. *betonica,* said by Pliny to have been first called *Vettonica* from the Vettones, a people of Spain,   B. officinalis, L.

„   WATER-, from similarity of leaf to that of the wood betony,   Scrophularia aquatica, L.

BIFOIL, the tway-blade, L. *bifolium,* two-leaf,

Listera ovata, R.B.

BIG, O.N. *bygg,* a name of Scandinavian origin applied in our northern counties and Scotland to the six-rowed variety of barley, and according to Holmboe (v. *bygg*), unquestionably related to *bua,* dwell, whence *byggia,* till, as being the grain most commonly cultivated in high latitudes,   Hordeum, L.

BIGOLD, tinsel, false gold, applied to a weed that is not the genuine *Golde,*   Chrysanthemum segetum, L.

BILBERRY or BULBERRY, Da. *böllebär,* dark-berry, from *böl,* dark,   Vaccinium Myrtillus, L.

BIND-WEED, a weed that binds,

Convolvulus arvensis, and sepium, L.

„   BLACK-,   Polygonum Convolvulus, L.

„ BLUE-, in Ben Jonson's Vision of Delight, the bitter-sweet,

"How the *blue* bindweed doth itself infold
With honeysuckle!"

Solanum Dulcamara, L.

BIND-WITH, a with used to bind up faggots, the Traveller's joy, Clematis Vitalba, L.

BIRCH, A.S. *beorc, birce,* or *byrc,* O.H.G. *piricha,* L.G. *barke,* Du. *berke,* Da. *birk,* Ic. and Sw. *biörk,* Russ. *bereza,* a *z,* as usual, replacing a German *k,* connected with Skr. *brichk,* Hindi *birchk,* a tree, and perhaps with Skr. *bhurgga,* a tree whose bark is used for writing upon. Klaproth argues from this word the northern origin of the dominant race in Hindostan, to whom this tree was the only one south of the Himalaya which they could name; all the others being new to them (Garnett, p. 33). It is the same word as *bark* in the two significations of tree-rind and vessel. In the first, as tree-rind, we find it forming A.S. *beorcan,* to bark, L.G. *bark, borke,* Eng. Du. and Da. *bark,* Ic. and O.N. *boerkr;* in the sense of vessel, the Lat. *barca,* which, as it stands isolated in its own language, is, no doubt, of foreign origin, the source of Fr. *barque,* Du. and Da. *barke,* Ic. *barkr,* Eng. *bark* and *barge.* In the earlier period of our Germanic race, while it was still confined to the northern latitudes, birch bark was used, as at the present day in the same countries, for boat building and roofing, and probably, as in Norway occasionally, for greaves for the legs, and from these different applications the tree took its name of bark-tree, a word connected with G. *bergen,* A.S. *beorgan,* protect, shelter, put into a place of safety. Betula alba, L.

BIRD'S BREAD, from the Fr. *pain d'oiseau,* the stonecrop, called so, apparently, from no better reason than its appearance in blossom when young birds are hatched,

Sedum acre, L.

BIRD-CHERRY, a cherry only fit for birds,
Prunus padus, L.

BIRD's-EYES, from its bright blue flowers,
Veronica Chamædrys, L.
and also, locally,   Primula farinosa, L.
  ,,   RED-,   Geranium Robertianum, L.

BIRD's-FOOT, from its incurved claw-like legumes,
Ornithopus perpusillus, L.

BIRD's-FOOT TREFOIL, from its legumes spreading like a
crow's foot,   Lotus corniculatus, L.

BIRD's-NEST, from its leafless stalks resembling a nest of
sticks, such as crows make,   Monotropa Hypopitys, L.
also, from its matted roots,   Neottia Nidus avis, L.

BIRD-SEED, canary grass used to feed birds,
Phalaris canariensis, L.

BIRD's-TONGUE, L. *lingua passerina*, from the shape of
the leaf,   Polygonum aviculare, L.

BIRTHWORT, from its supposed remedial powers in par-
turition, suggested, on the doctrine of signatures, by the
shape of the corolla, whence also its Greek name from
ἀριστος, best, and λοχεια, delivery,
Aristolochia Clematitis, L.

BISHOP's-LEAVES, from being known in French as *l'herbe
du siège*, in reference to its remedial powers in hemorrhoidal
affections, and this word *siège* being understood as of a
bishop's see,   Scrophularia aquatica, L.

BISHOP's-WEED, possibly from Fr. *levesque*, the name of
another umbelliferous plant, transferred to this, and mistaken as meaning bishop,   Ammi majus, L.

BISHOP's-WORT, a name applied, for reasons unknown,
to the Devil-in-a-bush,   Nigella damascena, L.

BISTORT, from its writhed roots, L. *bis*, twice, *torta*,
twisted, and thence called by Turner *Twice-writhen*,
Polygonum Bistorta, L.

BITTER CRESS,   Cardamine amara, L.

BITTER-SWEET, L. *Amara-dulcis*, from the rind of its stalk, which, as Turner observes b. iii. 2, " when it is first tasted is bitter, and afterwards sweet," a quality from which it gets its German name; *je lenger je lieber*.

Solanum Dulcamara, L.

„ also an apple so called, of which Gower says (ed. 1554, fol. 174):

> For all such time of love is lore,
> And like unto the *bittersuete*,
> For though it thinke a man fyrst suete,
> He shall well felen atte laste,
> That it is sower, and maie not laste.

Pyrus Malus, L. var.

BITTER VETCH, Vicia Orobus D.C.

BITTER-WORT, from the taste of the root, a name adopted from the German *bitterwurtz*, Gentiana, L.

BLACKBERRY, from the black colour of its fruit in contrast with that of the raspberry and dewberry,

Rubus fruticosus, L.

BLACK-BENT, -COUCH, or -SQUITCH, from its weedy character and dark purple flowers, the slender foxtail,

Alopecurus agrestis, L.

BLACK BINDWEED, from its want of the conspicuous white flowers of the other bind-weeds,

Polygonum Convolvulus, L.

BLACK BRYONY, from its dark-coloured glossy leaves, and black root, Tamus communis, L.

BLACK HOREHOUND, from its dingy colour in contrast with the white-leaved true horehound, Ballota nigra, L.

BLACK SALTWORT, in contrast with the Salsola,

Glaux maritima, L.

BLACK-SEED, the nonesuch, from its black head of legumes in contrast with the light yellow capitules of the hop-clover, Medicago lupulina, L.

BLACKTHORN, the sloe, from the conspicuous blackness

of its rind at the time of flowering, in contrast with that
of the white-thorn, 　　　　　　　Prunus spinosa, L.

BLACK-WORTS, whortle-berries, from their dark colour as
compared with the cow- and cranberry,

　　　　　　　　　　　　Vaccinium Myrtillus, L.

BLADDER CAMPION, from its inflated calyx,

　　　　　　　　　　　　Silene inflata, L.

BLADDER-FERN, from its bladder-like indusium,

　　　　　　　　　　　　Cystopteris fragilis, Bern.

BLADDER HERB, the winter cherry, which was supposed,
from its inflated calyx, to "cleanse the bladder, and open,
scour, and purge, the urinal passages,"

　　　　　　　　　　　　Physalis Alkekengi, L.

BLADDER-NUT, from its inflated seedpods,

　　　　　　　　　　　　Staphylea pinnata, L.

BLADDER-SNOUT, from the shape of the corolla, and
BLADDER-WORT, from the vesicles on its leaves,

　　　　　　　　　　　　Utricularia vulgaris, L.

BLEABERRY, or BLAEBERRY, from, *blae*, a word that in
our northern counties means "of a livid or pale-bluish
colour," Atk. Clevel. Dial. 　　Vaccinium uliginosum, L.

BLEEDING HEART, the name of the wallflower in the
western counties, more particularly the dark variety of it,
and apparently dating from a time when in its ordinary
state it was called *Heart's ease*,

　　　　　　　　　　　　Cheiranthus Cheiri, L.

BLESSED THISTLE, a thistle so called, like other plants
which bear the specific name of "blessed," from its supposed power of counteracting the effect of poison,

　　　　　　　　　　　　Carduus benedictus, L.

BLIND NETTLE, called so from the resemblance of its
leaves to those of a stinging nettle, and their not harming
or seeming to notice any body; whence in most languages
it bears a name that implies dead, deaf, or blind;

　　　　　　　　　　　　Lamium album, L.

BLINKS or BLINKING-CHICKWEED, the "*alsine flosculis conniventibus*" of Merret's Pinax, so called from its half-closed little white flowers peering from the axils of the upper leaves, as if afraid of the light, Montia fontana, L.

BLITE, L. *blitum*, Gr. βλιτον, insipid, the name of some potherb which Evelyn in his Acetaria takes to be the Good Henry, and remarks of it that " 'Tis insipid enough,"

Chenopodium Bonus Henricus, L.

in more modern works usually referred to   Blitum, L.

BLOOD-ROOT, from the red colour of its root and its consequent adoption, upon the doctrine of signatures, for the cure of the bloody flux. " Tormentilla in dysenteria quod rubra est." Linn. in Bibl. Botan. p. 117.

Tormentilla officinalis, L.

BLOOD-STRANGE, from *blood*, and a verb only found in composition, *strengen*, draw tight, and metaphorically, as in G. *harn-strange*, stop,   Myosurus minimus, L.

BLOOD-WORT, or BLOODY-DOCK, from its red veins and stems,   Rumex sanguineus, L.

BLOODY WARRIOR, from its crimson-tinged petals, the dark-blossomed wallflower,

Cheiranthus Cheiri, L.

BLOODY-MAN'S FINGER, from its lurid purple spadix,

Arum maculatum, L.

BLOW-BALL, the head of the dandelion in seed, from children trying their luck by blowing the pappus from its receptacle,   Leontodon Taraxacum, L.

BLUE-BELL, from the bell-shape of its flower,

Scilla nutans, Sm.

in Scotland   Campanula rotundifolia, L.

BLUE-BLAW, a name that would at first sight seem to be merely "blue blow or blossom," but the latter word is in old works invariably spelt *blaw*, and is rather the G. *blau*, blue, Prov. *blave*, of which the Fr. names of the flower, *blaveole*, *blavelle*, *blavet*, or *blaverolle* are the diminutives.

Blue-blaw is therefore a tautology, and means "blue-blue." Its Scotch name, *blawort*, is a better one.

Centaurea Cyanus, L.

BLUE-BOTTLE, from the bottle shape of the involucre, and its brilliant blue flower,   Centaurea Cyanus, L.

BLUE-CAPS, from its tuft of blue flowers,

Scabiosa succisa, L. and Knautia arvensis, Coult.

BOG ASPHODEL,   Narthecium ossifragum, Hud.

BOG-BEAN, either a translation of Fr. *trèfle des marais*, or a fancied correction of its proper name, *buck-bean*,

Menyanthes trifoliata, L.

BOG-BERRY, or BOG-WORT, the cranberry,

Vaccinium Oxycoccos, L.

BOG FEATHERFOIL, from its finely divided feathery leaves, or foils, the water violet,   Hottonia palustris, L.

BOG-MOSS,   Sphagnum, L.

BOG MYRTLE, or DUTCH MYRTLE, an evergreen aromatic shrub with some general resemblance to a myrtle, and abundant in peat-mosses,   Myrica Gale, L.

BOG-RUSH,   Schœnus, L.

BOG VIOLET, the butterwort,

Pinguicula vulgaris, L.

BOLBONAC, from the Arabic,   Lunaria biennis, L.

BOLT, Pr. Pm. *bolte*, petilium, tribulum,

Ranunculus, L.

BOODLE, see BUDDLE,   Chrysanthemum segetum, L.

BORAGE, Fr. *bourache*, M.Lat. *borago*, of which Apuleius says, that its former name was "*corrago*, quia cordis affectibus medetur," a word that the herbalists suppose to have become, by change of *c* to *b*, *borrago*. It is probably a Latinized Oriental name brought with the plant from Syria.   B. officinalis, L.

BORE-COLE, in Tusser BORE, according to Hettema, in Philol. Trans. 1858, from Du. *boerekool*, peasant cabbage,

Brassica oleracea, L. v. fimbriata.

BORE- or BOUR-TREE, the elder, "from the great pith in the younger branches, which children commonly bore out to make pop-guns," Ray, N. C. Words, but perhaps the O.N. *burr, baurr,* or *bör* (Holmboe), Sambucus nigra, L.

BOTTLE, see BLUE-BOTTLE.

BOTTLE-BRUSH, the field horsetail, from the resemblance of its barren frond to one, Equisetum arvense, L.

BOUTS, or BOOTS, the marsh marigold, from the Fr. *bouton d'or,* in respect of the yellow flower buds,
Caltha palustris, L.

BOWYER'S MUSTARD, from some apothecary probably,
Lepidium ruderale, L.

BOX, A.S. *box* and *bux,* L. *buxus,* Gr. πυξος, from πυξις, a pyx or turned box made of the wood,
Buxus sempervirens, L.

BOX-HOLLY, from its box-like leaves terminating in a prickle like those of holly, Ruscus aculeatus, L.

BOY'S-LOVE, or LAD'S-LOVE, the southernwood, from an ointment made with its ashes being used by young men to promote the growth of a beard: "Cinis Abrotani barbam quoque segnius tardiusque enascentem cum aliquo dictorum oleorum elicit." Matth. Comm. in Diosc. iii., 25; a purpose for which it is also recommended by Herbarius, c. ii. Artemisia Abrotanum, L.

BRAKES, from *brake,* G. *brache* or *brach-feld,* uncultivated land, a term used to replace the M.Lat. *fractitius* or *ruptitius ager,* land that is breakable, or again open to tillage after a term of years, land that is not preserved as forest. The fern so called is named from its place of growth in the same way as whin, heath, bent, and brier.
Pteris aquilina, L.

BRACKEN, supposed by Jamieson to mean female brake, but more likely to be a word introduced from Scandinavia and identical with Sw. *bräken,* which Rietz derives from Sw. *bräcka,* break. *Brägen,* in Aalborg, means, when applied to rye, a crop of that grain sown on land that has borne rye the previous year. Molb. D. D. Both these words seem, like *Brake*, to indicate what grows on fallow land.   Pteris aquilina, L.

BRAMBLE, A.S. *bremel, brembel,* or *brœmbel,* in Pr. Pm. *brymmeylle* or *brymbyll,* Du. *braam,* G. *brame,* O.H.G. *pramo,* words, which, as Grimm remarks, signify prickly or thorny bushes, but are connected etymologically with G. *brummen,* L. *fremere,* and others indicating "noise." Bramble means usually the blackberry bush,
     Rubus fruticosus, L.
but in Chaucer, l. 13676 :
     " The bramble flour that bereth the red hepe,"
is the dog-rose,   Rosa canina, L.

BRANDY-BOTTLE, a name usually explained as alluding to the odour of the flower, but rather more probably taken from the shape of the seed-vessel, the yellow water-lily, or can-dock,   Nuphar luteum, L.

BRANK, buckwheat, from a Latin word, *brance,* that occurs in Pliny, l. xviii. c. 7, where it seems rather to mean a barley : " Galliæ quoque suum genus farris dedere, quod illic *brance* vocant, apud nos sandalam, nitidissimi grani." The word will be identical with *blanc,* white, Port. *branco,* and equivalent to *wheat,* which properly means " white." See Diefenbach, Orig. Europ. p. 265. Pol. *pohanka.*   Polygonum Fagopyrum, L.

BREAKSTONE, from L. *saxifraga,* a plant that fissures a rock, understood as meaning a lithontriptic plant, to be administered in cases of calculus, a name applied to several different species belonging to different genera, viz.
     Pimpinella Saxifraga, L.   Alchemilla arvensis, D.C.
     and more particularly the genus Saxifraga, L.

BRIDE'S LACES, see LADY'S LACES.

BRIDEWORT, from its resemblance to the white feathers worn by brides,   Spiræa ulmaria, L.

BRIER, A.S. *brær*, Pr. Pm. *brere*, Fr. *bruyere*, called in Normandy *brière*, from the waste land on which it usually grows, M.Lat. *brugaria* or *bruarium*, W. *brueg*, a forest, Bret. *brûg*, from which *brugaria* would seem to have been formed, various wild species of   Rubus and Rosa,

   ,, SWEET-,   Rosa rubiginosa, L.

BRIER-ROSE, any wild rose, but chiefly the common hedge or dog-rose,   Rosa canina, L.

BRIMSTONEWORT, from its roots yielding, as W. Coles says, "a yellow sap which waxeth quickly hard, and dry, and smelleth not unlike to brimstone,"

          Peucedanum officinale, L.

BRINJAL, or BRINGALL, Port. *beringela*, from the Tamul *brinjaul*, its name in Ceylon and southern India,

          Solanum Melongena, L.

BRISTLE-FERN, from the bristle that projects beyond its receptacle,   Trichomanes radicans, Sw.

BROCCOLI, the plural of It. *broccolo*, a small sprout, diminutive of *brocco*, a shoot,

          Brassica oleracea, L. var. sabellica.

BROOK-LIME, in old writers BROK-LEMPE or -LYMPE,

          Veronica Beccabunga, L.

BROOK-WEED, from its growing beside brooks,

          Samolus Valerandi, L.

BROOM, A.S. *brom*, G. *brame*, a word of the same origin as *bramble*, and O.N. *brum*, foliage, but at present applied almost exclusively to a shrub of which besoms are made, and called from it *brooms*,   Spartium scoparium, L.

   ,, DYER'S-,   Genista tinctoria, L.

BROOM-RAPE, L. *rapum genistæ*, from *broom*, a plant, upon which it is parasitic, and *rape*, L. *rapa*, a turnip, which its clubby tuberous stem somewhat resembles,

          Orobanche, L.

BROWNWORT, A.S. *brunwyrt*, G. *braunwurz*, in Brunsfelsius and all the old herbalists spelt *brunnwurz*, said to be called so from the brown colour of its stems and flowers, but more probably from its growing so abundantly about the *brunnen* or public fountains of German towns and villages,   Scrophularia aquatica and nodosa, L. also, from its being supposed to cure the disease called in German *die braune*, a kind of quinsey, the Brunella, or, as it is now spelt, with a P,   Prunella vulgaris, L.

BRUISEWORT, from its supposed efficacy in bruises, the daisy,   Bellis perennis, L. and also the soapwort,   Saponaria officinalis, L.

BRYONY, L. *bryonia*, Gr. βρυωνια.

,,   BLACK-, from its dark glossy leaves and black root,   Tamus communis, L.

,,   WHITE-, from the paler colour of the leaves and of the root,   Bryonia dioica, L.

BUCKBEAN, believed by some botanists to have been originally *bog-bean*, which, from its French synonym, *tréfle des marais*, is very plausible, but that in Dutch also it is called *bocks-boonen*, and in German *bocksbohne*, and considered a remedy against the *scharbock*, or scurvy, whence it is called *scharbock's klee*. *Buckes-beane*, and not *bog-bean*, is the name of it in all the old herbals, and this must be admitted to be the proper and established one, being, no doubt, derived from the Dutch word, one which seems to be a corruption of L. *scorbutus*, the scurvy.   Menyanthes trifoliata, L.

,,   FRINGED-, so called from its delicately fringed corolla, and its alliance with the genuine buckbean,   Limnanthemum nymphæoides, Lk.

BUCK'S-BEARD, from its long coarse pappus, resembling the beard of a he-goat,   Tragopogon pratensis, L.

BUCK'S-HORNE, from its furcated leaves,   Plantago Coronopus, L.

BUCK-MAST, the nuts or mast of the beech, which was formerly called *bucke*,   Fagus sylvatica, L.

BUCKRAMS, from its offensive odour, see RAMSON.

Allium ursinum, L.

in Parkinson's Th. Bot. and some other Herbals, for *buck-rampe*, in allusion to the spathe and spadix,

Arum maculatum, L.

BUCKTHORN, from M. Lat. *spina cervina*, or *cervi spina* of Valerius Cordus, who seems to have misunderstood its German name *buxdorn*, i.e. *box-thorn*, a translation of the πυξακανθα of Dioscorides, to which this shrub and its congeners were referred by the earlier herbalists, for *bocksdorn*, the thorn of a buck, Rhamnus catharticus, L.

BUCKWHEAT, Du. *boekweit*, G. *buchwaitzen*, from the resemblance of its triangular seeds to beechnuts, a name adopted with its culture from the Dutch,

Polygonum Fagopyrum, L.

BUDDLE, in Tusser spelt BOODLE, Du. *buidel*, a purse, because it bears *gools* or *goldins*, gold coins, Du. *gulden*; a punning allusion to its yellow flowers so called;

Chrysanthemum segetum, L.

BUG AGARIC, a mushroom that used to be smeared over bedsteads to destroy bugs, Agaricus muscarius, L.

BUGLE, M.L. *bugula*, dim. of *abuga*, one of the various spellings of a word given by Pliny as corresponding to Gr. χαμαιπιτυς, and variously written *abiga, ajuga, iva*, etc. Ajuga reptans, L.

BUGLOSS, L. *buglossa*, from Gr. βους, an ox, and γλωσσα, tongue, descriptive of the shape and rough surface of the leaves, Anchusa officinalis, L.

  ,, VIPER'S-, Echium vulgare, L.

BUGLOSS COWSLIP, the lung-wort, from its having the leaves of a bugloss and the flowers of a primula,

Pulmonaria officinalis, L.

BULLACE, in the Grete Herball *bolays*, in Turner *bulles*, in Pr. Pm. *bolas*, the Sp. *bolas*, bullets, Lat. *bullas*, bosses on bridles, and called so from its hard round fruit.

               Prunus communis, Huds. var. insititia, L.

BULLOCK's LUNGWORT, from its curative powers in the pneumonia of bullocks, suggested, on the doctrine of signatures, by the resemblance of its leaf to a dewlap; see MULLEIN.   Verbascum Thapsus, L.

BULL-FIST, L. *bovista*,   Lycoperdon, L.

BULLS AND COWS, more commonly called "Lords and Ladies," the purple and the pale spadices, respectively, of

                        Arum maculatum, L.

BULL's-FOOT, from the shape of the leaf, the more commonly called coltsfoot,   Tussilago Farfara, L.

BULL-WEED, from O.E. *boll*, any globular body, such as the seed-vessel of flax, Dan. *bold*, a ball, It. *palla*, Gr. παλλα, the knapweed, so called from its globular involucre,   Centaurea nigra, L.

BULL-WORT, properly *pool-wort*, from its growing in or near pools,   Scrophularia, L.
also, in Gerarde, for the same reason,   Ammi majus, L.

BULRUSH, formerly spelt *pole-rush*, the pool-rush, *jonc d'eau*, A.S. *ea-risc*, from its growing in pools of water, and not, like the other rushes, in mire,   Scirpus lacustris, L.

BULL-SEGG, the *pool-segg* or *-sedge*, the reed-mace,

                        Typha latifolia, L.

BUMBLEKITES, the blackberry, from Scot. *kyte*, belly, as in the "Wife of Auchtermuchty":

    "The deil cut aff thair hands, quoth he,
      That cramm'd your *kytes* sae strute yestreen!"

and *bumble*, applied in Chaucer, l. 6554, to the voice of the bittern; from the rumbling and bumbling caused in the bellies of children who eat its fruit too greedily,

                      Rubus fruticosus, L.

BURDOCK, a name that, properly speaking, is a pleonasm; for *bur* and *dock* both meant originally, the one in French, and the other in the Germanic languages, a flock or lump of wool, flax, or hemp combed out in carding: but upon *dock* being extended from the *bur-dock* to other broad-leaved plants, the first syllable was added, to distinguish this species (which pre-eminently deserved the name by the trouble it gave housewives) from plants of the sorrel and other tribes. The word *bur* is the Fr. *bourre*, L. *burra*, and primarily means the lump or tangled mass of refuse fibre, of which the involucre of this species formed the nucleus. See HARDOCK. Arctium Lappa, L.

BUR MARIGOLD, a composite flower allied to the marigold, with seeds that adhere to the clothes, like burs,
Bidens tripartita, L.

BUR PARSLEY, from its bur-like bristly carpels,
Caucalis daucoides, L.

BUR REED, from its narrow reed-like leaves, and the burs formed by its seed vessels,
Sparganium ramosum, L.

BUR THISTLE, from its prickly involucre,
Carduus lanceolatus, L.

BUR-WEED, or BURDOCK CLOTWEED, a weed with large leaves and burs somewhat like those of the burdock or clotbur, Xanthium strumarium, L.

BURNET, a term formerly applied to a brown cloth, Fr. *brunette*, It. *brunetta*, dims. of *brun* and *bruno*, and given to the plant so called from its brown flowers,
Poterium Sanguisorba, L.

BURNET BLOOD-WORT, from its power of stanching blood, and its resemblance to burnet,
Sanguisorba officinalis, L.

BURNET SAXIFRAGE, from its supposed lithontriptic qualities, and the resemblance of its leaves to burnet,
Pimpinella Saxifraga, L.

BURSTWORT, from it supposed efficacy in ruptures,
Herniaria glabra, L.

BUTCHER'S BROOM, according to Loudon, and to the

authoress of "Sylvan Sketches," Mrs. Kent, from butchers making besoms of it to sweep their blocks ; according to Parkinson, Th. Bot. p. 254, to sweep their stalls. But it seems to have been called so from being used to preserve meat from rats and bats. Thus Tragus, ed. 1595, p. 340, "welchen dorn etliche *myacantham*, d.i. *murinam spinam* nennen, darumb das diser dorn der meusen und ratten zuwider ist." Matthioli, l. iv. c. 141, gives the same explanation of its Italian name, *Pungi-topi*, prick-rats : "Nos, quod arcendis muribus is aptissimus sit, si salitis appensis carnibus funiculo circumligetur, *Pungi-topi* Hetrusca lingua dicimus." So also Castor Durante in his Herbario nuovo, p. 402 : " Per avere i rametti vencidi ed le foglie dure, et horride, se ne fanno le scope e servono per cacciar, come si è detto, i topi e le nottole dalle carni salate." So also Lonicerus, Kraüterb., p. 204.

Ruscus aculeatus, L.

BUTCHER'S PRICKWOOD, from skewers being made of it,

Rhamnus Frangula, L.

BUTTER AND EGGS, from the colour of the flowers,

Linaria vulgaris, L.

BUTTERBUR COLTSFOOT, a plant so called perhaps from a confusion of Fr. *bourre*, whence *bur* in *burdock*, with Fr. *beurre*, butter. Dr. How and W. Coles derive the name from the leaves being used, as they suppose, for lapping butter in, a purpose to which they do not seem to be applied at the present day.

Petasites vulgaris, Dsf.

BUTTER-CUP, not, perhaps, from *butter* and *cup*, but rather more probably from Fr. *bouton d'or*, the bachelor's button, a name given to its double variety, the *cup* being the Old Eng. *cop*, a head, as in Wycliffe's Bible (Judg. ix. 7), a word that became obsolete, and was replaced by *cup*. It will have meant, originally, *button-head*. See GOLDCUP and KINGCUP.

Ranunculus, L.

BUTTER-DOCK, from its leaves being used for lapping butter, whence the Scotch name of it, *Smair-dock*,

Rumex obtusifolius, L.

BUTTER-FLOWER, Du. *boter-bloem*, G. *schmalz-blume*, from a mistaken notion that it gives butter a yellow colour; or, as Fuchs tells us, p. 878, from the greasy surface of the petals, Ranunculus, L.

BUTTER-JAGS, an obscure name, perhaps in the first place *bottle-jacks*, Lotus corniculatus, L.

BUTTERWORT, from its greasy feel, "as if," says W. Coles, "melted butter had been poured over it,"

Pinguicula vulgaris, L.

CABBAGE, Fr. *caboche*, It. *cabuccio*, from *cabo*, a head, being a variety of colewort that forms a round head,

Brassica capitata, L.

„ St. PATRICK'S, Saxifraga umbrosa, L.

„ SEA-, Crambe maritima, L.

CALAMINT, Gr. καλη, good, and μινθη, mint, as being a valuable antidote to the bite of serpents,

Calamintha officinalis, L.

CALATHIAN VIOLET, L. *Viola calathiana*, from L. *cala-thus*, Gr. καλαθος, basket, a name given by Pliny to some other very different plant, but by a mistake of Ruellius transferred to Gentiana Pneumonanthe, L.

CALE, COLE, or COLEWORT, A.S. *cawl*, and *cawlcyrt*, Du. *kool*, L. *caulis*, a stalk, a name given to a thick-stemmed variety, the kohl-rabi, and extended to the other kinds of cabbage. "The Apothecaries and common Herbarists do call it *caulis* of the goodnesse of the stalke," says Gerarde, p. 249. Brassica oleracea, L.

„ SEA-, Crambe maritima, L.

CALTROPS, A.S. *coltræppe*, in the romance of K. Alisander, 1, 6070, *calketrappen*, M.Lat. *calcitrapa*, from L. *calx*, a heel, and M.Lat. *trappa*, a snare, a name first applied to the caltrop used in war to impede the progress of cavalry, and thence to the spiny heads of this plant,

Centaurea Calcitrapa, L.

CALVES-FOOT, Fl. *calfsvoet*, Fr. *pied de veau*, from the shape of the leaf, Arum maculatum, L.

CALVES-SNOUT, the snapdragon, Fr. *mufle de veau*, from a fancied resemblance in the seed vessel. "Antirrhinon," says Cordus, "fructum fert vitulino capiti similem, tam exquisita similitudine, ut etiam os et nares appareant." He illustrates his statement with a caricature of the seed-vessel, which, as he gives it, certainly bears a most extraordinary likeness to a calf's skull.

Antirrhinum majus, L.

CALVERKEYS, in Awbrey's Wilts, probably the same as the *Culverkeys* of Walton's Angler, names no longer known.

CAMMOCK, A.S. *cammuc, -ec,* or *-oc,* a name given by old writers to two very different plants, a Peucedanum, and an Ononis. In the former sense it occurs in Piers Plowman's Vision, l. 13584:

"For communlike in contrees *Cammoke* and wedes
Foulen the fruyt in the feld, ther thei growen togideres."

In this sense of a Peucedanum it is found in several MSS. quoted by O. Cockayne in Leechdoms, v. ii., p. 374. The term *kambuck* is still given in Suffolk to the kexes.

Peucedanum officinale, L.

„ In the herbalists *cammock* means whin or rest-harrow, Ononis arvensis, L.

CAMPANELLE, It. *campanella*, a little bell, dim. of *campana*, a name given by Bulleyn to the hedge-bell,

Convolvulus sepium, L.

CAMPION, from having been used in the chaplets with which champions at the public games were crowned. It. *campione*, M.Lat. *campio*, from *campus*, a battle-field.

„ BLADDER-, from its inflated calyx,

Silene inflata, L.

CAMPION, CORN-, from its growing among corn,
Agrostemma Githago, L.

„ MEADOW-, the Ragged Robin,
Lychnis flos cuculi. L.

„ MOSS-, from its moss-like tufts,
Silene acaulis, L.

„ ROSE-, from the colour of the flower,
Lychnis coronaria, L.

CANARY-GRASS, from being a grass of the Canary Islands,
and used to feed Canary birds, Phalaris canariensis, L.

„ REED-, Digraphis arundinacea, Trin.

CAN-DOCK, from its broad leaves, and the shape of its
seed-vessel, like that of a silver can or flagon, Dan.
aa-kande, Nuphar luteum, Sm.

CANDLE-BERRY, from the fruit of an American species of
the genus yielding wax of which candles are made,
Myrica Gale, L.

CANDLE-RUSH, from its pith being used for rush-lights,
Juncus effusus, L.

CANDY-TUFT, or CANDY-MUSTARD, a tufted flower brought
from the Island of Candy, or Crete, Iberis umbellata, L.

CANKER, a tree-fungus, from its seeming to eat like a
cancer into a decaying tree, Boletus, L.

CANKER-ROSE, from its red colour and its detriment to
arable land, the field poppy, Papaver Rhœas, L.

CANTERBURY BELLS, so named by Gerarde, Ed. em.
p. 450, from growing very plentifully in the low woods
about Canterbury, Campanula Trachelium, L.

CAPER-PLANT, or WILD CAPER, from its immature seed
vessels being used in sauce for the buds of the real caper,
Euphorbia Lathyris, L.

CAPILLAIRE, the maiden-hair fern, from its being used
to prevent the hair from falling off, says Matthioli (l. iv.
c. 132.), quoting from Theophrastus : "ad defluvium capil-
lorum utile." Adiantum Capillus Veneris, L.

CAPON'S TAIL, from its spreading white flowers,
                              Valeriana officinalis, L.

CAPRIFOLY, M.Lat. *caprifolium,* goat's leaf, a mistake
for *capparifolium,* caper-leaf, the woodbine, which, in an
Anglo-Saxon translation of Dioscorides quoted by O. Cockayne (v. ii. p. 302), is spoken of as one " þe man capparis
and oþrum naman wudubend hateð," which is called *capparis,* and by another name *woodbine.* The similarity of
the leaf of this shrub to that of the caper, and its habit of
growing about walls and rocks, very naturally led the
northern nations to confuse them together, and the blundering mistake of *cappari* for *capri* has given rise to the Fr.
chevrefeuille, G. *geiss-blatt,* etc. Lonicera Caprifolium, L.

CARDOON, Fr. *cardon,* L. *cardunculus,* dim. of *carduus,* a
thistle,   Cynara Cardunculus, L.

CARLINE THISTLE, L. *Carolina,* so named after Charlemagne, Carl de groote, of whom the legend relates, as we
learn from Tabernæmontanus (vol. ii. p. 391), that " A
horrible pestilence broke out in his army, and carried off
many thousand men, which greatly troubled the pious
emperor. Wherefore he prayed earnestly to God, and in
his sleep there appeared to him an angel, who shot an
arrow from a cross-bow, telling him to mark the plant upon
which it fell, for that with that plant he might cure his
army of the pestilence. And so it really happened." The
herb thus miraculously indicated was this thistle.
                              Carlina vulgaris, L.

CARNATION, incorrectly derived in general from the flesh
colour of the flowers, and supposed to be connected with
L. *carne,* but more correctly spelt by our older writers
coronation, as representing the *Vetonica coronaria* of the early
herbalists, and so called from its flowers being used in
chaplets, *coronæ.* So Spenser, in his Shepherd's Calendar :
    " Bring *coronations* and sops in wine
    Worn of paramours."
                              Dianthus Caryophyllus, L.

CARNATION-GRASS, certain sedges, from the resemblance of their leaves to those of the carnation, more especially the Carex glauca, Scop., and C. panicea, L.

CARPENTER'S-HERB, from its corolla seen in profile being shaped like a bill-hook, and, on the doctrine of signatures, supposed to heal wounds from edged tools, the self-heal,
Prunella vulgaris, L.

CARRAWAY, M.Lat, *carui semina*, seeds of *careum*, Gr. καρον, Carian, so called from its native country, Caria. This genitive case was adopted for the name of the seed, as in Arundel MS. 42, f. 55, " Carui groweþ mykel in merys in þe feld, and in drye placys of gode erþe." Way's Pr. Pm. p. 333. Carum carui, L.

CARRAGEEN-MOSS, a sea-weed so called from an Irish word that means " a little rock," the name of some place in Ireland where it was first collected for sale,
Chondrus crispus, Lyngb.

CARROT, Fr. *carotte*, L. *carota*, Daucus Carota, L.

CARSE, an old spelling of *cress*, A.S. *cærs*.

CASE-WEED, or CASSE-WEED, so called in allusion to its little purse-like capsules, from Fr. *caisse*, L. *capsa*, a money-box, Du. *cas*, Capsella Bursa pastoris, L.

CASSIDONY, L. *stœchas sidonia*, from Sidon, where the plant is indigenous, Lavandula Stœchas, L.

CAT'S-EAR, from the shape of its leaves,
Hypochæris maculata, L.

CAT'S-FOOT, from its soft flower-heads,
Gnaphalium dioicum, L.
also, from the shape of its leaves, the ground ivy,
Nepeta Glechoma, Benth.

CATS-MILK, from its milky juice oozing in drops, as milk from the small teats of a cat,
Euphorbia helioscopia, L.

CAT-MINT, or CAT-NEP, " because," says Gerarde, p. 544, " cats are very much delighted herewith : for the smell of

it is so pleasant unto them, that they rub themselves upon
it, and wallow or tumble in it, and also feed on the branches
very greedily;" which singular statement the good old
herbalist copied from Dodoens (i. 4, 14), without, perhaps,
ascertaining its truth. Nepeta cataria, L.

CAT'S-TAIL, from its long cylindrical furry spikes,
Typha latifolia, L.
also from its cylindrical spike, Phleum pratense, L.

CATCH-FLY, from its glutinous stalks,
the genus Silene, and Lychnis viscaria, L.

CATCH-WEED, a weed that catches the passer by,
Galium Aparine, L.

CAULIFLOWER, L. of Bauhin's Pinax, *brassica cauliflora;*
of Parkinson, Par., p. 505, *caulis florida;* from L. *caulis,*
cole, and *flores,* flowers, formerly called *cole-flower, coley-
flowers,* and *cole-flourey,*
Brassica oleracea, L. var. florida.

CELANDINE, L. *chelidonium,* Gr. χελιδονιον from χελιδων,
swallow; "not," says Gerarde, p. 911, "because it first
springeth at the coming in of the swallowes, or dieth when
they go away, for, as we have saide, it may be founde all
the yeare, but because some holde opinion, that with this
herbe the dams restore sight to their young ones, when
their eies be put out;" an old notion quoted from Dodoens
(i. 2. 29), and copied by him from Pliny (l. xxv. c. i), and
by Pliny from Aristotle. This wonderful fact is received
and repeated by every botanical writer of those days, and
is embodied by the author of the Schola Salernitana, l. 217,
and by Macer, c. 52, in the couplet,

"Cæcatis pullis hac lumina mater hirundo,
(Plinius ut scripsit) quamvis sint eruta, reddit."

,, GREATER-, Chelidonium majus, L.
,, LESSER-, from its blossoming at the season when
the swallow arrives, the pilewort,
Ranunculus Ficaria, L.

CELERY or SELLERY, Fr. *celeri*, It. *sellari*, the plural of
sellaro, the name under which it was introduced in the seventeenth century, corrupted from L. *selinum*, Gr. σελινον,
Apium graveolens, L.

CENTAURY, or CENTORY, L. *centaurium*, a plant so called,
says Pliny, from the Centaur Chiron, who cured himself
with it from a wound he had accidentally received from an
arrow poisoned with the blood of the hydra. What plant
this was, is uncertain. The name is now given to the
knapweeds, which the Germans, resolving *centaurium* into
centum aureos, a hundred pounds, call *Tausend Gulden*.
Centaurea, L.

„ GREATER-, or MORE-, of the old black-letter herbals,
a gentianeous plant. Askham, in his Lytel Herball, says
of it, " It is named the More Centory or Earthgall: his
floures be yelowe in the croppe ;" and Dr. Linacre in his
version of Macer, that " More Centory or Earthgall hath
leves lyke to the Lesse Centory, but more whyter, and
yelowe flowers, and flowreth not but in the top."
Chlora perfoliata, L.

Lyte, and other herbalists since his time, incorrectly assign
the name to the knapweed,   Centaurea nigra, L.

„ LESSER-, so called in contrast with the greater- or
more- centory,   Erythræa Centaurium, L.

CENTINODE, or CENTYNODY, hundred knot, from its many
joints, L. *centum* and *nodus*,   Polygonum aviculare, L.

CETERACH, from a supposed Arabic or Persian word,
chetherak, which Stapel on Theophrastus (p. 1164) derives
from *pteryga*, corrupted to *peteryga* and *ceteryga*, and
meaning " winged," a doubtful etymology,
C. officinarum, J. Sm.

CHADLOCK, see CHEDLOCK.

CHAFE-WEED according to Hooker in Fl. .Lond., from
its use in Northumberland to prevent heavy loads from
galling the backs of beasts of burden ; but more probably

from its application to chafings of the skin in the human subject, or, as Ray expresses it in Cat. Plant. Cant., "quoniam ad intertrigines valet;" the cud-weed,

Gnaphalium germanicum, W.

CHAFF-WEED, A.S. *ceaf* and *weod*, from its small chaffy leaves, Centunculus minimus, L.

CHALOTS, see SHALLOT.

CHAMOMILE, L. *chamœmelum*, Gr. χαμαιμηλον, earth-apple, from the odour of its flowers,

Anthemis nobilis, L.

„ WILD-, Matricaria chamomilla, L.

CHAMPIGNON, Agaricus oreades, B.

CHANTARELLE, from the French, a mushroom so called, Cantharellus cibarius, Fr.

CHARLOCK, CARLOCK, or CALLOCK, in Scotland SKEL-LOCH, A.S. *cerlice*, which seems to be either formed from *cerre*, turn, and to indicate a vicarious plant, a weed of fallow ground; or, as is more probable, from L. *chœro-phyllum*, and thus connected with G. *schierling*. In a MS. of the fourteenth century in Rel. Ant. ii. 80, it is spelt *szerlock*, and translated *caroil*, chervil, which seems to confirm this view of its origin. The name is at present given to a wild mustard. Brassica Sinapistrum, Boiss.

CHECKERED LILY, from the markings on the petals, Fritillaria Meleagris, L.

CHEDDAR PINK, from its place of growth in Somerset-shire, on the cliffs of that picturesque ravine,

Dianthus cæsius, L.

CHEDLOCK, CHADLOCK, or KEDLOCK, A.S. *cedeleac*, from *leac*, a plant, and *cede*, which seems to be the same as L.Germ. *küdick*, *kettich*, *köddick*, Da. *kidike*, related, perhaps, to Da. *kiede*, annoy. In the eastern counties *chad* means the refuse sifted from wheat. The name is now confounded with *charlock*, but in Westmacott's Scripture Herbal, p. 86, and other old works, is assigned to the hemlocks. There is nothing related of St. Chad or Cedde, that in any way connects him with these weeds.

Brassica Sinapistrum, Boiss.

CHEESEBOULS, the red poppy, from the shape of the ripe capsule resembling that of round cheeses,

Papaver Rhœas, L.

CHEESE-RENNET, or -RUNNING, a name given to the yellow lady's bedstraw from its supposed power of curd-ling milk. "Galium inde nomen sortitum est suum, quod lac coagulet." Matth. l. iv. c. 91. A.S. *cys-gerun*, from *cys*, cheese, and *gerun*, a word connected with G. *rinnen* and *gerinnen*, Sw. *rönna*, coagulate.

Galium verum, L.

CHEET, the name of a spurious oil that in Gerarde's time was palmed off upon the public for the Spanish oil of Sesamum. See below, GOLD OF PLEASURE.

CHEIR, WILD-, the wallflower, from its Latin name,

Cheiranthus Cheiri, L.

CHEQUER-TREE, the service-tree, so called in Evelyn's Sylva, and in Sussex at the present day, from *Choker*, the choke-pear, being an antique pronunciation of the word which we find in the humorous old ballad of The Frere and the Boy, l. 115,

> " Whan my fader gyveth me mete,
> She wolde theron that I were *cheke*." *i.e.* choaked.

See CHOKE-PEAR. Sorbus domestica, L.

CHERRY, from O.E. CHERISE, as it is spelt in Chaucer, a word that was mistaken for a plural of *cherry*, Fr. *cerise*, It. *ciriegia*, L. *cerasca*, adj. of *cerasus*, Gr. κερασος, a name brought with the tree from Asia Minor,

Prunus Cerasus, L.

,, BIRD'S-, a sort fit for birds only,

Prunus Padus, L.

,, WILD-, or Gean, Prunus avium, L.

CHERRY-WOOD, in Jacob's Pl. Faversh. the water-elder,

Viburnum Opulus, L.

CHERVIL, A.S. *cærfille*, Fr. *cerfeuil*, L. *chærophyllum,*
Gr. χαιρεφυλλον, from χαιρω, rejoice, and φυλλον, leaf,
Chærophyllum sylvestre, L.

„ HEMLOCK-, or ROUGH-,
Caucalis Anthriscus, L.

CHESSES, a name that by some mistake has been transferred to this plant, the peony, from the poppy, which, from
the shape of its capsule, was called *chasses* and *chese-boules*,
Pæonia corallina, L.

CHESTNUT, L. *castanea*, Gr. καστανον,
Castanea vesea, Lam.

„ HORSE-, probably so named from its coarseness,
but according to Parkinson (Th. Bot., p. 1402), from being
used in Turkey as a food for horses suffering from shortness
of wind, Æsculus Hippocastanum, L.

CHEVISAUNCE, in Spenser's Sheph. Cal. April, l. 142:
"The pretty pawnce, And the *chevisaunce*,"
evidently a misprint for *cherisaunce*, comfort, heart's-ease,
the *cheiri* or wallflower, the plant to which the name of
Heart's-ease was originally given. The word is omitted in
the glossaries to Spenser, but occurs in Chaucer's Romaunt
of the Rose, l. 3837:
"Then dismayed I left all soole,
Forwearie, forwandred as a foole,
For I ne knew ne *cherisaunce*."
Cheiri is the Moorish name *Keiri*, with which the plant
now so familiar to us was brought hither from Spain.
Cheiranthus Cheiri, L.

CHICK-PEA, or CHICHES, It. *cece*, L. *cicer*,
C. arietinum, L.

CHICKLING, a spurious *Chick*; cf. Vetchling and Crambling; Lathyrus, L.

CHICKWEED, A.S. *cicena-mete*, from the various plants
comprehended under this name having been used to feed
chickens. "It is thought to be wholesome for sick birds,

whence called *Chickweed."* Threlkeld. " On en consomme
beaucoup pour la nourriture des oiseaux de volière."
Duchesne, s. l. plantes utiles, p. 226.

Stellaria media, L.

also, in Hudson,   Veronica arvensis, and agrestis, L.

  „   MOUSE-EAR-,   Cerastium vulgatum, Huds.

  „   WINTER-GREEN-,   Trientalis europæa, L.

CHICORY, L. *Cichorium,* Gr. κιχορη or κιχοριον, an
Egyptian word. " Intybum in Ægypto Cichorium vocant."
Plin. N.H. l. xx. c. 8.   C. Intybus, L.

CHIER, WILD-, see CHEVISAUNCE, the wallflower, from
an Arabic word, *Keiri,*   Cheiranthus Cheiri, L.

CHILDING CUDWEED, a parturient cudweed,

Filago germanica, L.

CHILDING PINK, a parturient pink, one that is called so
from its throwing out younger and smaller flowers like a
family of little children round it. *Childing* is an expression
analogous to ' calving,' ' kittening,' etc., and occurs frequently in old authors. Thus in Lev. xii. 3, Wycliffe's
version has, "If a woman *childiþ* a male child:" and in
Gen. iv. 1, 2, " Eve *childide* Cain, and eft sche *childide* his
brother Abel."   Dianthus prolifer, L.

CHIVES, in R. Turner's Bot. p. 175, CIVES, Fr. *cives,*
derived by Diez from L. *cepa,*

Allium Schœnoprasum, L.

CHOKE-PEAR, Fr. *poire d'estranguillon,* L. *pyrum strangu-
latorium,* Ger. p. 1270, a name given to a wild pear so hard
and austere as to choak ; with an allusion, perhaps, to the
death of Drusus, a son of the Emperor Claudius, which
was caused by a fruit of this character,

Pyrus communis, L.

CHRISTOPHER, see HERB CHRISTOPHER, a name given to
several different plants.

CHRISTMAS, from being used for decoration at that season,
the holly,   Ilex Aquifolium, L.

CHRISTMAS ROSE, from its open rose-like flower, and its blossoming during the winter months,

Helleborus niger, L.

CHRIST'S LADDER, an old name, for we find it as *Christis leddere* in catalogues of the fourteenth century. From the plant having been called *fel*-wort, earth-*gall*, *fel*-terræ, etc., we may suspect that it has arisen from *Christis galle*, Christ's gall, or *Christis schale*, Christ's Cup, having been mistaken for *Christi scala*, Christ's ladder, and that it alludes to the bitter draught offered to Jesus upon the cross. Erythræa Centaurium, L.

CHURL'S HEAD, from its rough hairy involucre,

Centaurea nigra, L.

CHURL'S TREACLE, garlick, from its being regarded as a *Triacle* or antidote to the bite of venomous animals. See POOR MAN'S TREACLE. Allium sativum, L.

CHURNSTAFF, from its straight stem spreading into a flat top, Euphorbia helioscopia, L.

CIBBOLS, Fr. *ciboule*, It. *cipolla*, scallions,

Allium Ascalonicum, L.

CICELY, Gr. σεσελι, some umbelliferous plant.

  „ ROUGH-, Caucalis Anthriscus, Huds.

  „ SWEET-, from its agreeable odour,

Myrrhis odorata, L.

  „ WILD-, Chærophyllum sylvestre, L.

CIDERAGE, Fr. *cidrage*, Polygonum Hydropiper, L.

CINQUEFOIL, in A. Askham's Lytel Herball *Quynckefolye*, Fr. *cinq* and *feuilles*, L. *cinque foliola*, so called from its five leaflets, Potentilla, L.

  „ MARSH-, Comarum palustre, L.

CISS, abbreviated from *Cicely*.

CLAPPEDEPOUCH, a nickname meaning *clap*- or *rattle-pouch*, from *clap*, Du. *klappen*, a name that alludes to the licensed begging of lepers, who stood at the cross-ways with a bell and a clapper. Hoffmann von Fallersleben in his Niederländische Volkslieder says of them, p. 97, "Separated from all the world, without house or home, the lepers were obliged to dwell in a solitary wretched hut by the road-side; their clothing so scanty that they often had nothing to wear but a hat, and a cloak, and a begging wallet. They would call the attention of the passers-by with a bell or a clapper, and receive their alms in a cup, or a bason at the end of a long pole. The bell was usually of brass. The clapper is described as an instrument made of two or three boards, by rattling which they excited people to relieve them." The lepers would get the name of Rattle-pouches, and this be extended to the plant in allusion to the little purses which it hangs out by the way-side. Capsella Bursa pastoris, L.

CLARY, M.Lat. *sclarea*, a word formed from *clarus*, clear, by prefixing the preposition *ex*, whence It. *schiarire* and *schiarare*. This word *Clary* affords a curious instance of medical research. It was solved by the apothecaries into *clear-eye*, translated *Oculus Christi*, *Godes-eie*, and *See-bright*, and eye-salves made of it. See Gerarde, p. 627.

Salvia Sclarea, L.

  ,, WILD-,   Salvia Verbenaca, L.

CLAVER, Du. *klaver*, the old and correct way of spelling *Clover*. See CLOVER.

CLEAVERS, or CLIVERS, the goosegrass, A.S. *clife*, Du. *kleef-kruid*, from its cleaving to the clothes, or possibly from Da. *klyve*, O.N. *klifa*, climb, O.Fris. *klieve*. It is likely that in this, as in so many other cases, a word, understood in one county in one sense, has been adopted, with some slight change, in another county in a different but equally appropriate sense; or that one form of the word has been learnt from a Dutch or Flemish book, and the other from a Friesic or Scandinavian. Galium Aparine, L.

CLIDERS, see CLITE.

CLIFF-PINK, from its growing upon Cheddar Cliffs in Somersetshire, Dianthus cæsius, L.

CLITE, CLITHE, CLIDERS, and CLITHEREN, a name of the goosegrass, probably from *Cliver*, by a change, not unfrequent, of *v* to *th*; or from Du. *klederen*, G. *kleider*, clothes; see CLEAVERS. Galium Aparine, L.

CLIMBERS, from its habit of *climbering*, or attaching itself to objects, the Fr. *grimper*, originally identical with *griper*, clutch, a use of the word found in Tusser, p. 109,

> " Set plenty of boughs among runcival pease,
>   To *climber* thereon, and to branch at their ease."

the Virgin's bower, Clematis Vitalba, L.

CLOG-WEED, a shortened form of *keyc-logge*, as it is spelt in Turner of Tottenham, quoted by Way in a note to Pr. Pm. p. 278, a word formed of *keck*, a hollow stalk, and *lock*, A.S. *leac*, a plant, and signifying the *kex-plant*,

Heracleum Sphondylium, L.

CLOSE SCIENCES, the Dame's Violet, called, as Parkinson tells us (Th. Bot. p. 628), the single variety of it, *Single Sciney*, and the double variety, *Close Sciney*, from which Gerarde made this ridiculous name. *Sciney*, no doubt, has arisen from its specific name, *Damascena*, understood as *Dame's Scena*. Hesperis matronalis, L.

CLOT-BUR, the bur-dock, called in Chaucer and in Pr. Pm. CLOTE, sometimes spelt incorrectly CLOD-BUR, A.S. *clatte*, G. *klette*, a bur that sticks to clothes, a word connected with many others beginning with *cl* or *kl*, such as *cleave, climb, cling, kletten*. The name may possibly have some connection with *clout*, through a confusion between the Latin name of the plant, *lappa*, and Du. *lap*, G. *lappen*, a clout, whence our English verb *lap*; as in a line of Pierce Plowman:

> "Thereon lay a litel chylde *lapped* in *cloutes*."

Arctium Lappa, L.

CLOUD-BERRY, from its growing on the cloudy tops of mountains, Ger. p. 1368, Rubus Chamæmorus, L.

CLOVE GILLIFLOWER, from its scent of clove, Sp. *clavo*, a nail, the shape of the spice so called,

Dianthus Caryophyllus, L.

CLOVER, or, as it is more correctly spelt in all the herbals, and all our older writers, and in Lowland Scotch, CLAVER, Du. *klaver*, A.S. *clæfer* and *clæfra*, Da. *klever*. It is evidently a noun in the plural number, probably a Frisian word, and means "clubs," from L. *clava*, and refers to the *clava trinodis* of Hercules. It is in fact the *club* of our cards, Fr. *trèfle*, which is so named from its resemblance in outline to a leaf with three leaflets. Trifolium, L.

,, ALSIKE-, Trif. hybridum, L.

,, BIRDSFOOT-, from its claw-like legumes,

Lotus corniculatus, and Trigonella ornithopodioides, L.

,, CRIMSON-, Trif. incarnatum, L.

,, DUTCH-, Trif. repens, L.

,, HARESFOOT-, from its furry soft capitules,

Trif. arvense, L.

,, HARTS-, Melilotus officinalis, L.

,, HEART-, from the markings of the leaf,

Medicago maculata, L.

,, HOP-, from the shape of its fruiting capitules,

Trif. agrarium, and procumbens, L.

,, HORNED-, of W. Turner, the lucerne,

Medicago sativa, L.

,, MEADOW-, Trif. pratense, L.

,, STRAWBERRY-, from the shape of its capitules, when in fruit, with the calcyces pink and inflated,

Trif. fragiferum, L.

CLOWN'S ALLHEAL or WOUNDWORT, so called by Gerarde, p. 852, because a countryman, who had cut himself to the bone with a scythe, healed the wound with it in seven days, Stachys palustris, L.

CLOWN'S LUNG-WORT, from its use in pulmonary disease,
                          Lathræa Squamaria, L.
CLUB-MOSS, a mossy plant with a club-like inflorescence,
                          Lycopodium, L.
CLUB-RUSH, from its club-like inflorescence, the reed-
mace,   Typha latifolia, L.
COB-NUT, from *cob*, a thick lump, A.S. *copp*, head, so
called from being used in a game called *cob-nut*,
                          Corylus Avellana, L. var. grandis.
COCK'S COMB, of botanists, from the shape of the calyx,
                          Rhinanthus Crista galli, L.
COCK'S COMB, of Hill's, and some other herbals, from the
shape of its legume, the sainfoin,   Onobrychis sativa, L.
COCK'S COMB of the gardeners,   Celosia cristata, L.
COCK'S-COMB-GRASS, from the shape of the panicle,
                          Cynosurus echinatus, L.
COCK'S-FOOT, from the shape of the spike,
                          Dactylis glomerata, L.
COCK'S-HEAD, from the shape of the legume, the sainfoin,
                          Onobrychis sativa, L.
COCKS, from children fighting the flower-stems one
against the other. See KEMPS. Plantago lanceolata, L.
COCKLE, A.S. *coccel*, L. *caucalis*, Gr. καυκαλις, some
umbelliferous plant, which Clusius says (p. ccii.) was the
same as δαυκος ἀγρια.   *Cockle* or *Cokyl* was used by
Wycliffe and other old writers in the sense of a weed
generally, but in later works has been appropriated to the
gith, or corn pink.   Agrostemma Githago, L.
CODLIN, originally *coddling*, from *coddle*, to stew or boil
lightly, a boiling apple, an apple for coddling or boiling,
a term used in Shakspeare, (T. N. i. 5,) of an immature
apple, such as would require cooking to be eaten, but now
applied to a particular variety,   Pyrus Malus, L. var.
CODLINS AND CREAM, from the odour of its flowers,
or of its fresh shoots, or according to Threlkeld, of its

bruised leaves, the larger willow-herb so named after a once favourite dish alluded to in King's Cookery:
"In cream and codlings rev'ling with delight."

Epilobium hirsutum, L.

COL, abbreviated by the Apothecaries from *Coliander* once used for *Coriander*,   Coriandrum sativum, L.

COLESEED, or COLLARD; see CALE; rape,

Brassica Napus, L.

COLEWORT, or COLLET, cabbage,   Brassica oleracea, L.

COLMENIER, a name given in the Herbals to the Sweet William, and also spelt *Tolmeiner*, which in Parkinson is divided into *Toll-me-neer*, as though the meaning had been *Cull me-*, or *Toll me near*, probably a fanciful explanation of a name derived from some unknown foreign word, *d'Allemagne*, perhaps; see TOLMEINER.

Dianthus barbatus, L.

COLT'S-FOOT, L. *ungula caballina*, from the shape of the leaf,   Tussilago Farfara, L.

COLTZA, Flem. *kool-zaad*, cole-seed,

Brassica Napus, L.

COLUMBINE, L. *columbina*, adj. of *columba*, pigeon, from the resemblance of its nectaries to the heads of pigeons in a ring round a dish, a favourite device of ancient artists,

Aquilegia vulgaris, L.

COMFREY, L. *confirma*, from its supposed strengthening qualities,   Symphytum officinale, L.

  ,, SPOTTED-, see LUNGWORT.

CONSOUND, or CONSOUD, L. *consolida*, "quia tanta præstantia est, ut carnes, dum coquuntur, conglutinet addita, unde nomen:" Pliny, xxvii. 6. a name given in the middle ages to several different plants, and among them to the daisy,   Bellis perennis, L.
to the comfrey,   Symphytum officinale, L.
to the bugle,   Ajuga reptans, L.
and to the wild larkspur,   Delphinium Consolida, L.

CONVAL LILY, L. *lilium convallium,* lily of combes, incorrectly translated " Lily of the *valley.*" The expression is used in the Vulgate translation of the Bible (Cant. ii. 1), and is appropriately given to this plant, as the flower of hollows surrounded by hills, its usual place of growth, although certainly not the flower meant by the royal poet.

Convallaria majalis, L.

COP-ROSE, from it red rose-like flower and the cop- or button-like shape of its capsule, Papaver Rhœas, L.

CORAL-ROOT, from its branching and jointed root-stock resembling white coral, Corallorhiza innata, R.B.

CORAL-WORT, from its white root, and the "divers small round knobs thereon resembling the knaggy eminences of coral," W. Coles, p. 56, Dentaria bulbifera, L.

CORD-GRASS, called so by Turner, because he " saw that rishe in the islands of East Friesland, and the people there make *ropes* of that rishe, and thache their houses also wyth the same," Spartina stricta, Sm.

CORIANDER, a plant, says Cogan, p. 26, "commonly called Coliander," Gr. κοριαννον, of κορις, a bug, from its odour, Coriandrum sativum, L.

CORK, the orchil, Norw. *korkje,* a corruption of an Arabic word into one more familiar,

Roccella tinctoria, D.C.

,, in the Highlands, Lecanora tartarea, Ach.

CORMEILLE, CORR, or CARMYLIE, Gael. *caermeal,* the heath-pea, a word adopted from the Gaelic,

Orobus tuberosus, L.

CORN, one of several words, which being common to widely separated branches of the Ind-European race, prove the practice of tillage among our ancestors before they left their first home in central Asia, Skr. *karana,* Go. *kaurn,* L. *granum,* Russ. *zerno,* a term applied to the several kinds of grain most commonly used in their respective countries. Max Müller refers the word to Skr. *kurna,* ground.

CORN-BIND, see BIND-WEED and BEAR-BIND.

CORN-BOTTLE, see BLUE BOTTLE.

CORN-COCKLE, see COCKLE.

CORN-FLOWER, from its being one of the gayest and most conspicuous wild flowers of corn-fields, Du. *korenbloem,* Latin in Ort. San. *Flores frumentorum,*

Centaurea Cyanus, L.

CORN-HONEWORT, from its use in curing the hone, or boil on the cheek,   Petroselinum segetum, L.

CORN-MARIGOLD, see MARIGOLD,

Chrysanthemum segetum, L.

CORN-PINK, in Northamptonshire (Baker), the corn-cockle,   Agrostemma Githago, L.

in some counties,   Campanula hybrida, L.

CORN-POPPY, or -ROSE,   Papaver Rhœas, L.

CORN-SALAD,   Valerianella olitoria, L.

CORN-VIOLET,   Campanula hybrida, L.

CORNEL, It. *corniolo.* L. *corneolus,* dim. of *corneus,* adj. of L. *cornus,* Gr. κρανεια,   Cornus sanguinea, L.

CORNISH-MONEYWORT, from its round leaves, and its having been first discovered in Cornwall, and long supposed to be peculiar to that county, and called there *Penny-pies,*   Sibthorpia europæa, L.

CORONATION, the older and more correct spelling of *carnation,* from its M.Lat. name *Vettonica coronaria,* as in Dodoens, ii. 1, 18, Tabern. vol. ii. c. 1, and Lyte, b. ii. ch. vii, who, in speaking of Clove Gillofers, says, "Tho greatest and bravest sort of them are called *coronations* or *cornations.*" See CARNATION. Dianthus Caryophyllus, L.

COSTMARY, L. *costus amarus,* its name in Bauhin's Th. Bot. p. 674, Fr. *coste amere,* misunderstood as *Costus Mariæ,* from Gr. κοστος, some aromatic plant unknown, an error that has very naturally arisen from this one having been dedicated to St. Mary Magdalene, and called after her, *Maudlin,* either in allusion to her box of scented

ointment, or to its use in the uterine affections over which
she presided. In old authors it occurs as *Herba sanctæ* or
divæ Mariæ. See MAUDLIN. Balsamita vulgaris, W.

COTTON-GRASS, or -RUSH, a grass-like plant with seed-
spikes resembling tufts of cotton from the protrusion of
the hypogynous bristles beyond the glumes,
Eriophorum polystachyum, L.

COTTON-WEED, from its soft white pubescence,
Gnaphalium, L.

COUCH-GRASS, or QUITCH, A.S. *cwice*, from *cwic*, vivacious, L. Germ. *quek, quik,* or *queek,* on account of its
tenacity of life, for, as says the Bremen Glossary (v. iii.
p. 401), " Kein gewächs hat mehr lebenskraft als der
Queck, wie die land- und garten-bauer mit verdruss er-
fahren," a name given to several creeping grasses, but
more especially to Triticum repens, L.

COUGH-WORT, a translation of G. βηχιον, a name given
to it from its medicinal use, the colts-foot,
Tussilago Farfara, L.

COVENTRY BELLS, from their abundance near that town
(Ger. em. p. 448), Campanula Trachelium, L.

COVENTRY RAPES, see RAMPION, called *rape* from its
tuberous turnip-like roots, Campanula Rapunculus, L.

COWBANE, from its supposed baneful effect upon cows,
Cicuta virosa, L.

COWBERRY, apparently from a blunder between *Vaccinium,* the fruit of the whortle, and *vaccinum,* what belongs
to a cow, Vaccinium Vitis idæa, L.

COW-CRESS, a coarse cress, Lepidium campestre, L.

COWSLIP, A.S. *cuslyppa* or *cusloppe,* which O. Cockayne
(Leechdoms, ii. p. 378), would derive from *cu,* cow, and
slyppa, slop, an explanation of it which is by no means
probable. Still less so is Wedgwood's, who, in ignorance
that cows do not eat cowslips, would regard the last syllable as a corruption of *leek.* There is very little poetry in

these popular names, and the word seems really to allude to a very humble part of dress. In the Stockholm medical M.S. it is spelt *kousloppe,* and evidently means "hoseflap," from Flem. *kouss,* hose, and *lopp,* flap. Such a name could scarcely have been given in the first place to the plant now called "cowslip," but was very applicable to the large oval flannelly leaf of the mullein, from which it has been transferred to it through the Latin name *Verbascum,* which comprehended both the cowslip and the mullein. This view is confirmed by its French name *braiette.*

                                         Primula veris, L.

,,   FRENCH- or MOUNTAIN-,   P. Auricula, L.

,,   JERUSALEM-,   Pulmonaria officinalis, L.

COW'S LUNGWORT, see BULLOCK'S LUNGWORT.

COW-PARSLEY, or COW-WEED,

                                Chærophyllum sylvestre, L.

COW-PARSNEP,   Heracleum Sphondylium, L.

COW-QUAKE, a word that would seem to have arisen from a confusion in German works between *queck* lively, a name given to the couch grasses, and *quäg,* cattle. Thus Bauhin tells us (Th. Bot. p. 9), that *Queckgras,* quitch, is so called by the Saxons [the people of Lower Germany] from the cattle being fond of it : "a jumentis quæ ea herba delectantur ; *Queck* enim ipsis 'jumentum' significat." The quaking grass is ranged by Tragus under these *queck-gräser,* and the similarity of G. *queck* and E. *quake* has fixed the name upon this species. The four words are in fact identical, etymologically speaking, and mean "alive ;" whence their various applications, as *queck, quitch, quake,* and *quäg,* to objects in which life and motion are conspicuous.   Briza media, L.

COW-WEED, see COW PARSLEY.

COW-WHEAT, from its seed resembling wheat, but being worthless as food for man,   Melampyrum, L.

CRAB, Sc. *scrab,* from A.S. *scrobb,* a shrub, implying a

bush- or wild-apple, in the Grete Herball called a "Wood-crabbe," and according to Turner (b. ii. p. 47), "in the north countre a *Scarb-tre,*" Fr. in Cotgrave, *pomme de boys,*   Pyrus Malus, L.

CRAB-GRASS, from its growing on the sea-shore, where crabs abound, and being supposed to afford them food,
     Salicornia herbacea, L.

CRAKE-BERRY, the crow-berry, O.N. *kraka,* a crow, whence Da. *krake-bär,* from its black colour, or, according to Dr. Johnston, in East. Bord. from crows eating it greedily,   Empetrum nigrum, L.

CRAMBLING ROCKET, a spurious *crambe,* or mustard (as vetchling is a spurious vetch), with the leaves of rocket,
     Sisymbrium officinale, L.

CRANBERRY, from its fruit being ripe in the spring, when the crane returns, Da. *tranebœr,* from *trane,* a crane, a name of late introduction, for Lyte calls them Marrish Whorts and Fenberries, and says (b. vi. c. 11) that "there is none other name for them known,"
     Vaccinium Oxycoccos, L.

CRANE'S-BILL, from the form of the seed vessel,
     Geranium, L.

CRAP, or CROP, buck-wheat, related to L. *carpere. Crop* in our old writers means a head of flowers, a cyma, and may have been given to this plant, as being thus distinguished from the cereal grains, which have no such conspicuous flowers. There is probably some prefix lost from the word.   Polygonum Fagopyrum, L.

CRAPPE, in some works, for no obvious reason, applied to the ray-grass,   Lolium perenne, L.

CRAZY, or CRAISEY, in Wiltshire and the adjoining counties, the buttercup, apparently a corruption of *Christ's eye,* L. *oculus Christi,* the medieval name of the marigold, which, through the confusion among old writers between *caltha* and *calendula,* has been transferred to the marsh

marigold, and thence to other ranunculaceæ. Thus in
M.S. Sloane, No. 5, *Oculus Christi* is explained "calendula, solsequium, the Seynt Marie rode ;" and again,
Gesner explains *Caltha*, which usually means the marsh
marigold, "ringel-bluom, *solsia*, quod solem sequatur,
vulgo *calendula*, quasi *calthula*. See MARYBUD.

Ranunculus, L.

CRESS, G. *kresse*, Fr. *cresson*, It. *crescione*, M. Lat. *cris-
sonium*, derived by C. Stephans, and by Diez, from L. *cres-
cere*, grow, "a celeritate crescendi." The form of the
word now in use has probably been adopted from the
Netherlands with the cultivation of the plants. Used absolutely, it means the genus   Lepidium, L.

   ,, BELLEISLE-,   Barbarea præcox, RB.
   ,, BITTER-,   Cardamine amara, L.
   ,, GARDEN-,   Lepidium sativum, L.
   ,, LAND-, in distinction from the water-cress,
Barbarea vulgaris, RB.
   ,, PENNY-, from its round silicules,
Thlaspi arvense, L.
   ,, SCIATICA-, from its medicinal use,   Iberis, L.
   ,, SWINE'S-, or WART-,   Coronopus Ruellii, D.C.
   ,, TOWN-, from its cultivation in *tounes* or gardens,
Lepidium sativum, L.
   ,, WALL-, from its usual place of growth, Arabis, L.
   ,, WATER-,   Nasturtium officinale, L.
   ,, WINTER-,   Barbarea vulgaris, RB.

CROCUS, Gr. κροκος, saffron.

CROSS OF JERUSALEM, from the resemblance of its
scarlet flower, both in shape and colour, to a Maltese or
Jerusalem cross,   Lychnis chalcedonica, L.

CROSS-FLOWER, from its flowering in Cross-week. See
ROGATION FLOWER.   Polygala vulgaris, L.

CROSS-WORT, from its cruciate or cross-placed leaves,
Galium cruciatum, Scop.

CROW-BELLS, the daffodil,

                Narcissus Pseudonarcissus, L.

CROWBERRY, from the black colour of its fruit,

                Empetrum nigrum, L.

CROW-FLOWER, the buttercup, from the resemblance of its leaf to a crow's foot. See CROWFOOT.

                Ranunculus acris and bulbosus, L.

but in old authors oftener applied to the Ragged Robin,

                Lychnis flos cuculi, L.

CROWFOOT, from being supposed, from the shape of its leaf, to be the Coronopus or crow's-foot of Dioscorides,

                Ranunculus, L.

CROWFOOT CRANESBILL, a geranium with a leaf like that of a crowfoot,   Geranium pratense, L.

CROW-GARLICK, a worthless one,   Allium vineale, L.

CROW-LEEKS,   Scilla nutans, Sm.

CROW-NEEDLES, or CRAKE-NEEDLES (Ray) from the long beaks of the seed vessels,   Scǎndix Pecten, L.

CROW-TOES, from its claw-like spreading legumes,

                Lotus corniculatus, L.

CUCKOO'S BREAD, or CUCKOO'S MEAT, or GOWK-MEAT, from its blossoming at the season when the cuckoo's cry is heard, M.Lat. of Ort. San. c. xviii, *panis cuculi*,

                Oxalis Acetosella, L.

" CUCKOO BUDS of yellow hue," Shaksp. (L.L.L. v. 2), are probably the buds of the crowfoot.

CUCKOO FLOWER, a name given in old works to the ragged robin,   Lychnis flos cuculi, L.

but now more generally to the lady's smock, which, as Gerarde says, p. 203, " flowers in April and May, when the cuckoo doth begin to sing her pleasant notes without stammering,"   Cardamine pratensis, L.

CUCKOO GILLIFLOWER, one of the plants formerly comprehended under the gilliflowers, and blossoming at the time of the cuckoo's song,   Lychnis flos cuculi, L.

CUCKOO-GRASS, a grass-like-rush flowering at the time of
the cuckoo, Luzula campestris, L.

CUCKOO-PINT, or -PINTLE, from A.S. *cucu*, lively, and
pintle (see Bailey), L.Ger. *pintel*, Fris. *pint* and *peynth*,
words explained in Outzen's Glossary, a plant so called
from the shape of the spadix, and its presumed aphrodisiac
virtues, and, although the editor of "Saxon Leechdoms"
(ii. p. 337) may not acquiesce in this derivation, and may
choose to overlook its synonyms, Wake Pintle, Wake
Robin, and others, most certainly not so called, in the first
place at least, after the cuckoo; although in later times it
may, through carelessness, and ignorance of the true derivation, have come to take that meaning; for, pace tanti
viri, how could such a name be given to it after a bird?
See below, WAKE-PINTLE and WAKE-ROBIN.
Arum maculatum, L.

CUCUMBER, Fr. *concombre*, It. *cocomero*, L. *cucumis, -eris*,
C. sativus, L.

CUDBEAR, from a Dr. Cuthbert Gordon, who first manu-
factured a dye from it,
Lecanora tartarea, Achar.

CUCKOO SORREL, A.S. *geaces sure*, the wood-sorrel, from
its flowering at the season when the cuckoo sings,
Oxalis Acetosella, L.

CUDWEED, from *cut*, Du. *kutte*, A.S. *cwið*, vulva, a plant
that on account of its soft cottony pubescence was used to
prevent chafings or to relieve them (see CHAFEWEED).
Gnaphalium germanicum, and uliginosum, L.
,, SEA-, Diotis maritima, L.

CULL-ME-, CUDDLE ME-, or CALL ME TO YOU, see PANSY.

CULLIONS, It. *coglione*, augm. of *coglia*, L. *coleus*, from
its double tubers, Orchis, L.

CULRAGE, through the French from L. *culi rabies*, a
plant so named, says Gerarde (p. 361) "from his operation
and effect when it is used in those parts." See Lobel,

Adv. Nov. p. 134.   Piers of Fulham says :
> " An erbe is cause of all this rage
> In oure tonge called *culrage*."
>
> Polygonum Hydropiper, L.

CULVERKEYS, a name found in Walton's Angler, and the same probably as *Calverkeys*, in Awbrey's Wilts, one now no longer used or understood. Being applied to a meadow plant it cannot be, as supposed by the commentators, the columbine, but far more probably, as suggested by Mr. Edw. King, in Notes and Queries, 2nd s. vii, 303, the blue-bell or common hyacinth.   Scilla nutans, Sm.

CULVERFOOT, in Lupton, (b. ix. No. 15), the doves-foot cranesbill,   Geranium columbinum, L.

CULVERWORT, A.S. *culfre*, pigeon, and *wort*, from the resemblance of its flowers to little heads of such birds feeding together, the columbine,   Aquilegia vulgaris, L.

CUMMIN, from Ar. *al qamoun*, Cuminum cyminum, L. .

CUP in Butter-cup, King-cup, and Gold-cup, not from a drinking vessel, but from the resemblance of its double variety to the gold head of a button, A.S. *copp*, a stud, Fr. *bouton d'or*,   Ranunculus acris, L.

CUP LICHEN, or CUP-MOSS, from its cup-like shape,
            Scyphophorus pyxidatus, Hook.

CURRANT, a name transferred from the small grape brought from Corinth, and thence called *Uva Corinthiaca*, to the fruits of several species of Ribes,
            R. rubrum and nigrum, L.

CUSHION-PINK, from its dense tufted growth, and the resemblance of its flowers in their general appearance to pinks,   Statice Armeria, L.

CUT-HEAL, the valerian, which was probably so called from its supposed efficacy in uterine affections, Du. *kutte*, A.S. *cwið*, but in mistaken conformity to its name, used, as Gerarde tells us, "in sleight cuts, wounds, and small hurts."   Valeriana officinalis, L.

CYCLAMEN, an adopted Latin name, Gr. κυκλαμινος.

C. europæum, L.

CYDERACH, the culrage. See CIDERAGE and CUDWEED.

CYPHEL, an unexplained name, possibly the Gr. κυφελλα, a mass of clouds, from its growth on cloud-capped Alpine heights,   Cherleria sedoides, L.

CYPRESS, L. *cupressus*, Gr. κυπαρισσος,

C. sempervirens, L.

CYPRESS-ROOT, or SWEET CYPRESS, from L. *cyperus*, a plant the aromatic roots of which are known as *English galingale*,   Cyperus longus, L.

DAFFADOWNDILLY, DAFFODILLY, AFFODILLY, and DAFFO-DIL, L. *asphodelus*, from which was formed *Affodilly*, the name of it in all the older writers, but subsequently confused with that of another flower, the so-called *sapharoun-* or saffron *lily*.

> "The thyrde *lylye* gyt there ys,
> That ys called felde lylye, y wys,
> Hys levys be lyke to *sapharoun*,
> Men know yt therby many one."

MS. Sloane, No. 1571.

With the taste for alliteration that is shown in popular names, the *Sapharoun-lily*, upon blending with *affodilly*, became, by a sort of mutual compromise, *daffadown-dilly*, whence we get our *daffodilly* and *daffodil*. This explanation of it is merely conjectural, and wants the test of historical evidence, but appears to be the best. The dictionaries derive it from "fleurs d'affodille;" but there is no such name to be found in any work, French or English, and it is highly improbable that a plant should be called. the "flowers" of the plant. Neither does this explain the *-down-* of *Daffadowndilly*. Narcissus Pseudonarcissus, L.

DAISY, A.S. *dæges-eage*, eye of day, O.E. *Daieseyghe*, from its opening and closing its flower with the daylight,

a name that seems to have delighted Chaucer, who makes
long and repeated allusions to it,   Bellis perennis, L.

,,   GREAT-, or MOON-, or OX-EYE-,
              Chrysanthemum Leucanthemum, L.

DAMASK VIOLET, or DAME'S VIOLET, L. *Viola Damascena*, from Damascus in Syria, Fr. *Violette de Damas*, misunderstood for *Violette des dames*,
                        Hesperis matronalis, L.

DAMSONS, DAMASINS, or DAMASK PRUNES, Fr. *damascene*,
a kind of plum first brought from Damascus,
              Prunus communis, Huds. var. damascena.

DANDELION, Fr. *dent de lion*, lion's tooth, L. *leontodon*,
a name, about the meaning of which modern authors are
undecided.   Some derive it from the whiteness of the root;
some from the yellowness of the flower, which they compare to that of the heraldic lion, whose teeth are of gold;
most of the Herbalists from the runcinate jags of the leaf,
which somewhat resembles the jaw, but certainly not a
tooth of the lion; others from other grounds more or less
plausible, but all to the neglect of the only safe guide in
these matters, the ancient writer who gave the name.
We learn from the Ortus Sanitatis, ch. 152, that a Master
William, who was a surgeon, and who seems, from ch. 226,
to have written a "cyrorgi," or work on surgery, was very
fond of this plant on account of its virtues, and therefore
likened it to a lion's tooth, called in Latin *dens leonis*.
"Diss kraute hat Meyster Wilhelmus, eyn wuntartzet
gewest, fast lieb gehabt umb seiner tugent willen, unnd
darumb hatt er es geglichen eynem leuwen zan, genant za
latein *dens leonis*."   Ed. Augsburg. 1486, fol.   It bears a
similar name in nearly every European language.
                        Taraxacum officinale, Vill.

DANEWORT, DANEWEED, or DANESWEED, names of the
dwarf elder for which Awbrey in his Nat. Hist. of Wilts,
p. 50, substitutes that of DANESBLOOD, and gives an ex-

planation of it that seems to be a fanciful one, seeing that
the plant bears the same name of "Danesweed" in other
counties, viz. that it grows in great plenty about Slaughterford, where there was a great fight with the Danes. Parkinson (Th. Bot. p. 208) derives it with more probability
from the aptness of the plant to cause a flux called the
Danes; but as the plant is expressly recommended by
Platearius as a remedy " contra *quotidianam,*" the *Dane*
may be a corruption of the last syllables of this word.

Sambucus Ebulus L.

DAPHNE, the name of a nymph, who was turned into a
shrub by the other gods, when pursued by Apollo,

D. Laureola, L.

DARNEL, in Pr. Pm. DERNEL, a name that in old writers
did not mean exclusively the large ray grass to which
we now assign it, but many other plants also, of many
different genera and natural orders, leguminosæ, gramineæ,
caryophylleæ, etc. The most probable source of this, as
of most other popular names, is its medical use. We find
that it was a specific remedy for cutaneous diseases.
Glantvilla (Batman's translation, 1582) says, c. 194, that
"Ray medled with brimstone and with vinegar helpeth
against scabs wet and dry, and against tetters, and against
itching." Now these diseases were called *zerna;* "Zernam
medici impetiginem vocant," says Cassius Felix, as quoted
by the editor of Macer on the line, descriptive of its
virtues :

" *Zernas* et lepras cura compescis eadem."

It is from this word that we seem to have got *dernel,*
which, so far as it was a specific name, meant "itch-weed."
But, in truth, there was great confusion among our early
herbalists in respect to the names of their plants, and
under that of *Darnel* were comprehended all kinds of corn-
field weeds. So in the Grete Herball, ch. 246, we find
under the picture of a vetch (!), " Lolium is Cokyll." The

A.S. version of Matth. ch. xiii. v. 25, renders the Lat.
" zizania" *coccel,* Wycliffe's both *cokel* and *darnel,* and later
versions *tares,* and Th. Newton, in his Herbal to the Bible,
p. 226, tells us expressly that, " under the name of *Cockle*
and *Darnel* is comprehended all vicious, noisome, and un-
profitable graine, encombring and hindering good corne."
The explanation given above is the most plausible that
offers itself, but the origin of this word is extremely
obscure, and all analysis of it quite conjectural. Some
incidental notice may another day throw a light upon it,
that cannot be elicited by any amount of thought, or inge-
nuity of conjecture.   Lolium temulentum, L.

DAUKE, the wild carrot, L. *daucus,* Gr. δαυκος, a word
that seems to be etymologically identical with the northern
laukr, leac, lauch, by a replacing of *d* with *l,*
D. Carota, L.

DEADMAN'S FINGERS, from the pale colour and hand-like
shape of the palmate tubers,   Orchis maculata, L.

DEAD NETTLE, Lat. of Ort. San. *Urtica mortua,* a plant
that has nettle-like leaves, but does not resent the touch
with a sting, and from its apparent insensibility is called
dead, deaf, and blind,   Lamium, L.
      ,,   ,, WHITE-,   L. album, L.
      ,,   ,, RED-,   L. purpureum, L.
      ,,   ,, YELLOW-,   L. Galeobdolon, Cr.

DEAD-TONGUE, from its paralysing effects on the organs
of voice, of which Threlkeld gives a striking example, on
the authority of a Mr. Vaughan, in the case of eight lads
who had eaten it, and of whom " five died before morning,
not one of them having spoken a word,"
Œnanthe crocata, L.

DEADLY NIGHTSHADE, or DEATH'S HERB,
Atropa Belladonna, L.

DEAF NETTLE, the Dead nettle,   Lamium, L.

DEAL-TREES, the species of fir that produce the deal of commerce,   Pinus and Abies.

DEER'S HAIR, from its tufts of slender stems looking like coarse hair,   Scirpus cæspitosus, L.

DELT-ORACH, an orach whose leaves are triangular, like a Greek letter *Δ*,   Atriplex patula, L.

DEPTFORD PINK, from its growing, according to Gerarde, "in a field next Deptford, as you go to Greenwich,"

                   Dianthus Armeria, L.

DEVIL IN THE BUSH, from its horned capsules peering from a bush of finely divided involucre,

                   Nigella damascena. L.

DEVIL'S-BIT, G. *Teufels abbiss*, L. *Morsus diaboli*, so called, says the Ortus Sanitatis, c. cclxi, on the authority of Oribasius, " because with this root the Devil practised such power, that the mother of God, out of compassion, took from the Devil the means to do so with it any more ; and in the great vexation that he had that the power was gone from him, he bit it off, so that it grows no more to this day." Threlkeld records a legend, that " the root was once longer, until the Devil bit away the rest, for spite ; for he needed it not to make him sweat, who is always tormented with fear of the day of judgment." Later authors explain it, as though the root would cure all diseases, and that the Devil, out of his inveterate malice, grudges mankind such a valuable medicine, and bites it off.   Scabiosa succisa, L.

DEVIL'S DARNING NEEDLES, from its long awns,

         ·   Scandix Pecten, L.

DEVIL'S GUTS, from the resemblance of the stem to cat-gut, and the mischief it causes, the dodder,   Cuscuta, L.

DEVIL'S MILK, from its acrid poisonous milk,

                   Euphorbia, L.

DEW-BERRY, G. *tauben-beere*, Norw. *col-bär*, supposed to be called so from the dove colour of its fruit, A.S. *duua*,

Du. *duif*, a dove, but perhaps with better reason referrible to the *theve-thorn* of Wycliffe's Bible.   See THEVE-THORN.
Rubus cæsius, L.

DEW-GRASS, from its rough dew-besprent blades, the cocksfoot grass,   Dactylis glomerata, L.

DEWTRY, from L. *Datura* (see Hudibras, iii. c. 1),
D. Stramonium, L.

DILL-SEED, from O.N. *dilla*, lull, being used as a carminative to cause children to sleep,
Anethum graveolens, L.

DITCH-BUR, called by Turner *Dychebur*, from its bur-like involucre, and its growth on *dykes*, not in *ditches*, as its modern name would lead us to suppose, the dyke being the dry bank that confines the water,
Xanthium strumarium, L.

DITTANDER, or DITTANY, apparently a corruption of L. *dictamnus*, the name of a very different plant, but applied to a cress, of which Lyte says (b. v. ch. 66), " It is fondly and unlearnedly called in English Dittany. It were better in following the Douchemen to call it Pepperwurt."   Lepidium latifolium, L.

DOCK, A.S. *docca*, which seems to be the same word as Norw. *dokka*, G. *docke*, Dan. *dukke*, a bundle of flax or hemp, a word that corresponds to Fr. *bourre*, a flock, and O.E. *harde* or *herde*, explained by Batman on Bartholomew (c. 160), as " what is called in Latin *stupa*, and is the clensing of hempe or flexe."   The name *Dock* would seem from this to have been first given to the burdock from the frequent occurrence of its involucres entangled in wool, and to have been transferred from this to other broad-leaved plants.   Used absolutely, it means at the present day those of the sorrel kind.   See BURDOCK and HARDOCK.
Rumex, L.

  „   BUR-, see under BURDOCK.
  „   CAN-, see under CANDOCK.

DOCK, FIDDLE-, from the shape of its leaves,

Rumex pulcher, L.

,, ROUND-, the common mallow, still called so in the charm that is used by children who have been stung with nettles, and alluded to by Chaucer in Troilus and Cressida (b. iv. sb. 62) : " Nettle in, dock out."

Malva sylvestris, L.

,, SHARP-, the sorrel, from its acidity,

Rumex Acetosa, L.

,, VELVET-, Verbascum Thapsus, L.

DODDER, the plural of Fris. *dodd*, a bunch, Du. *dot*, hampered thread, from its resemblance to bunches of threads entangled in the plants on which it grows,

Cuscuta, L.

DOGBERRY, or DOG-CHERRY, the fruit of the *Dogwood* tree, misunderstood as referring to the quadruped. See DOGWOOD. Cornus sanguinea, L.

DOG'S CHAMOMILE, a spurious or wild kind,

Matricaria Chamomilla, L.

DOG-GRASS, called so, Tabernæmontanus tells us, " sintemal sich die Hunde, wenn sie Massleid haben, damit purgiren ;" and R. Turner (Bot. p. 89) " It is called in Latin *gramen caninum*, because dogs eat the grass when they are sick." Triticum caninum, Huds.

DOG'S MERCURY, or DOG'S COLE, a spurious kind, to distinguish it from the so-called English Mercury,

Mercurialis perennis, L.

DOG'S ORACH, a stinking kind,

Chenopodium olidum, Sm.

DOG'S PARSLEY, G. κυναπιον, a worthless weed, parsley for a dog, Æthusa Cynapium, L.

DOG-ROSE, κυνορροδος and κυνοσβατος, a wild kind, so called from its want of scent and beauty, Rosa canina, L.

DOG'S-TAIL-GRASS, from its spike being fringed on one side only, Cynosurus cristatus, L.

Dog's-tongue, a translation of L. *cynoglossum*, a name given to it from its soft leaf,   C. officinale, L.

Dog's-tooth-grass, Fr. *chien dent*, from the sharp-pointed shoots of its underground stem,

Triticum caninum, Hud.
and Cynodon Dactylon, R.

Dog-Violet, a scentless one,   Viola canina, L.

Dog-wood, not so named from the animal, but from skewers being made of it. "It is rather a shrub than a tree," says Threlkeld, "the dry wood wonderfully resists the axe and the wimble, and is used for skewers by the butchers." *Dog*, in this view of it, is the Fr. *dague*, It. and Sp. *daga*, Fl. and Old Engl. *dagge*, equivalent to G. *dolch*, a *dagger*, and A.S. *dalc* or *dolc*, a fibula, a brooch-pin, and related to Skr. *dag*, strike. The verb *dawk* is still retained in the Western counties in a Nursery rhyme: "Prick it and *dawk* it, baker's man." This derivation of the name is supported by its synonyms *Prick-wood*, *Skewer-wood*, and *Gadrise*, but has been overlooked, and the fruit, from a mistaken idea of its meaning, called a *Hound's-berry*.   Cornus sanguinea, L.

Doob-grass, the name given in India to the dog's-tooth-grass,   Cynodon Dactylon, R.

Dove's foot, from the shape of the leaf,

Geranium molle, L.

Drake, Drawk, or Dravick, Du. *dravig*, W. *drewg*, Br. *draok*, darnel, cockle, or weeds in general, L. *daucus*, with insertion of *r*, as in Sp. *tronar* from *tonar*. It is sometimes found spelt *drank*, a form of the word which seems to have arisen, in the first place, from a mere misprint of *n* for *u*.   Bromus sterilis, and Avena fatua, L. etc.

Dropwort, according to Turner (b. iii. 31), from its small tubers hanging by slender threads,

Spiræa Filipendula, L.

DROPWORT, WATER-, from its use in stillicidium, and growth in wet places, Œnanthe fistulosa, L.

DRY-ROT, a name given to several species of fungus destructive of wood, which they render dry and as if destroyed by fire (see Proceedings of Linn. Soc. for 1850, p. 80), but probably derived from *tree*, wood, A. S. *treow*, and *rot*, . Merulius lacrymans, Wulf.

DUCK-MEAT or -WEED, an aquatic plant, a favourite food of ducks, called in Pr. Pm. *ende-mete*,
Lemna minor, L.

DULSE, Gael. *duillisg*, from *duille*, leaf, and *uisge*, water, a name given to several species of rose-spored algæ, and more especially to
Rhodomenia palmata, and Iridæa edulis, B. St. V.

DUNSE-DOWN, a pleonasm, from Du. *dons*, which means down, so called from its soft spikes, but whimsically derived by Lobel from its making people *dunch* or deaf, if it gets into their ears (Kruydtb. p. 113), the reed-mace,
Typha latifolia, L.

DUTCH CLOVER, or simply DUTCH, from the seed of it having been very largely imported from Holland, 150 tons annually, says Curtis in his Flor. Lond.
Trifolium repens, L.

DUTCH MYRTLE, L. *Myrtus Brabantica*, from its abounding in Dutch bogs, and replacing the myrtle of more genial climates, Myrica Gale, L.

DUTCH RUSH, a rush-like plant imported from Holland,
Equisetum hyemale, L.

DWALE, Da. *dwale*, torpor, trance, whence *dwale-bær*, a dwale- or trance-berry. In Chaucer (1. 4159), it is used for a sleeping draught: "There nedeth him no *dwale*;" and in Lupton's 1000 notable things, we have (b. iv. 1) a receipt for making *Dwale* for a patient to take, "while he be cut, or burned by cauterising," the ingredients of which are the juices of hemlock, nep, lettuce, poppy, and henbane, mixed up with pig's gall and vinegar. Once a general term, it has been appropriated to the deadly nightshade.

Atropa Belladonna, L.

DYER'S GREEN-WEED, in the sense of a dye-herb that tinges green,   Genista tinctoria, L.

DYER'S ROCKET OR YELLOW WEED, from its leaves resembling those of the genuine rocket, and its being used by the dyers to dye woollen stuffs yellow,

Reseda Luteola, L.

EARTH-BALLS, truffles, balls that grow under the earth,

Tuber cibarium, Sib.

EARTH-GALL, A.S. *eorð-gealle*, from their bitterness, plants of the gentian tribe, more particularly the lesser centaury,   Erythræa Centaurium, L.

EARTH-MOSS,   Phascum, L.

EARTH-NUT, or -CHESTNUT, ERNUT, or YERNUT, from its nutty esculent tubers,   Bunium flexuosum, With.

EARTH-SMOKE, L. *fumus terræ*, see FUMITORY,

Fumaria officinalis, L.

EARTH-STAR, a fungus so called from its stellate shape when burst and lying on the ground,   Geaster, Berk.

EGG-PLANT, from the shape of its fruit,

Solanum Melongena, L.

EGLANTINE, a name that has been the subject of much discussion, both as to its exact meaning, and as to the shrub to which it properly belongs. In Chaucer and our other old poets it is spelt *Eglantere* and *Eglatere*, as in a passage in the Flower and Leaf, st. 3 :

The hegge also, that yede in compas,
And closed in all the greene herbere,
With sicamour was set and *Eglatere.*

But whether this word meant originally the sweetbrier, the yellow rose called in systematic works Eglanteria, the dog-rose, the burnet rose, or some other species, cannot now be ascertained, and perhaps the poets who used it meant no more than a rose of any kind indifferently. Indeed, Milton in the expression " twisted eglantine," is supposed to have meant the woodbine. The derivation of the name is obscure. Diez, the highest authority in questions of French etymology, derives it from Lat. *aculeus*, a prickle, through *aculentus*, whence O. Fr. *aiglent*, covered with prickles, and *aiglentier*, which became *églantier*, and *églantine*, and in this view is supported by Em. Egger. The name seems in ancient French works to have been given to the wild roses. In our own early writers, and in Gerarde and the herbalists, it was a shrub with white flowers that was meant. At the present day by *Eglantine* is usually understood the sweet-brier, Rosa rubiginosa, L.

ELDER, A.S. *ellen* and *ellarn*, in Pierce Plowman *eller*, words that seem to mean "kindler," and to be derived through A.S. *œld*, Da. *ild*, Sw. *eld*, fire, from *œlan*, kindle, and related to Du. *helder*, clear, whence *op-helderen*, kindle or brighten up, a name which we may suppose that it acquired from its hollow branches being used, like the bamboo in the tropics, to blow up a fire, Sambucus nigra, L.

,, DWARF-, Sambucus Ebulus, L.

,, WATER-, Viburnum Opulus, L.

ELECAMPANE, L. *Enula campana*, by countrymen, says Isidore, called *Ala campana*, the latter word from its growing wild in Campania, the former from L. *Inula*, a word of uncertain derivation, Inula Helenium, L.

ELEVEN O'CLOCK LADY, Fr. *dame d'onze heures*, from its waking up and opening its eyes so late in the day,
Ornithogalum umbellatum, L.

ELF-DOCK, the elecampane, from its broad leaves called a *dock*, and from some confusion between its Italian name, *ella*, and the Dan. *elle*, an elf, deriving its prefix,
Inula Helenium, L.

ELM, a word that is nearly identical in all the Germanic and Scandinavian dialects, but does not find its root in any

of them. It plays through all the vowels, Ic. *Almr*, Da.
Alm, Ælm, and *Elm*, A.S. and Engl. *Elm*, Germ. in different dialects *Ilme, Olm*, and *Ulme*, Du. *Olm*, but stands
isolated, as a foreign word, which they have adopted. This
is the Lat. *Ulmus*, the terminating syllable of which, *mus*,
indicates an instrument, a material, or means, with which
something is done; while the first seems to be the *ul* of
ulcus, sore, and *ulcisci*, punish, in allusion to the common
use of rods of elm for whipping slaves. See Plautus (Asin.
2, 2, 96). The foreign origin of the name indicates that
the tree was introduced into England from the South of
Europe, and explains what Aubrey remarks in his "Wilts,"
that in the Villare Anglicum, although there are a great
many towns named after other trees, there are only three
or four *Elme*-tons. Ulmus, L.

EMONY, the anemony misunderstood as *An Emony*.
"Gardeners commonly call them Emonies." R. Turner.
Bot. p. 18. Anemone, L.

ENCHANTER'S NIGHTSHADE, a name that, by some
blunder, has been transferred from the mandrake, Atropa
Mandragora, to an insignificant garden weed. The mandrake was called *Nightshade* from having been classed with
the *Solana*, and *Enchanter's* from its Latin name *Circæa*,
Gr. κιρκαια, given to it after the goddess Circe, who be-
witched the companions of Ulysses with it (Od. b. x.); or
according to Dioscorides, as quoted by Westmacott, p. 105:
"'Twas called Circæa, because Circe, an Enchantress ex-
pert in herbs, used it as a Tempting-powder in amorous
concerns." C. lutetiana, L.

ENDIVE, It. and Sp. *endivia*, L. *intybea*, adj. of *intybus*,
Cichorium Endivia, L.

ENGLISH MERCURY, a plant reckoned among the *Mer-
curies*, but why called *English* more particularly, we are
not told, Chenopodium Bonus Henricus, L.

ERS, the bitter vetch, Fr. *ers*, L. *ervum*, E. Ervilia, L.

ERYNGO, L. *eryngium*, G. ἡρυγγιον, from ἐρυγγανειν, eructare, being, according to the herbalists, a specific against that inconvenience, E. maritimum, L.

EUPHRASY, Milton (P. L. b. xi. l. 414), Stockholm Med. M. S. *Ewfras*, Off. L. *euphrasia*, Gr. εὐφρασια, cheerfulness, a name that C. Bauhin tells us was first given to the bugloss, meaning probably the borage, and which has been transferred to the eyebright,

Euphrasia officinalis, L.

EVENING PRIMROSE, from its pale yellow colour and its opening at sunset, Œnothera biennis, L.

EVERLASTING FLOWER, from its retaining shape and colour when dried, Gnaphalium, L.

EVERLASTING PEA, Fr. *pois eternel*, from not being, like the common and the sweet pea, an annual,

Lathyrus latifolius, L.

EYE, the pink, in Tusser called "Indian eye," L. *ocellus*, as in the name of the carnation, *ocellus Damascenus*, from the eye-shaped marking of the corolla, Dianthus, L.

EYEBRIGHT, so called, as W. Coles tells us in his Adam in Eden, from its being used by the linnet to clear its sight, and thence adopted by men. Where he picked up this tale, does not appear. It is one of several similar legends. See CELANDINE, HAWKWEED, and PIGEON'S GRASS. We may add to this number of eye remedies learnt from the birds, that the use of fennel was taught to man by serpents:

Hac mansa serpens oculos caligine purgat:
Indeque compertum est humanis posse mederi
Illam luminibus, atque experiendo probatum est.
                                        Macer. c. xvii. l. 4.

We are told the same of the eagle, in reference to the use of the wild lettuce: "Dicunt aquilam quum in altum volare voluerit, ut prospiciat rerum naturas, Lactucæ sylvaticæ folium evellere, et succo ejus sibi oculos tangere, et

maximam inde claritudinem accipere." Apuleius, c. **xxx.**
The plant was long in vogue as a remedy in diseases of the
eye. Brunschwygk tells us in his quaint old German :
" Es was ouch ein küngin in Engelant, die brant allein das
wasser uss der blümlin, und thett wunderliche ding darmit
zû der ougen, als mir der selbigen küngin artzet geseyt
hat." Euphrasia officinalis, L.

EYEBRIGHT COW-WHEAT, a plant in some respects resembling both the eyebright and the cow-wheat,
Bartsia Odontites, Hud.

FAIR MAIDS OF FEBRUARY, snowdrops, from their white
blossoms opening about the 2nd of that month, when
maidens dressed in white walked in procession at the feast
of the Purification, Galanthus nivalis, L.

FAIR MAIDS OF FRANCE, a double-flowered variety of
crowfoot introduced from France,
Ranunculus aconitifolius, L.

FAIRY BUTTER, a fungus so called in the northern
counties from its being supposed to be made in the night,
and scattered about by the Fairies. Atkinson Clev. Dial,
p. 169. Tremella arborea and albida, Sm.

FAIRY FLAX, a flax so called from its delicacy,
Linum catharticum, L.

FALLEN STARS, from their sudden appearance glittering
on gravel walks after a night's growth, certain gelatinous
fungi, and particularly Tremella Nostoc, L.

FANCY, an attempted explanation of *Pansy,*
Viola tricolor, L.

FAT HEN, G. *Fette Henne,* a name given in Germany
and by the older herbalists to the orpine, Sedum telephium,
called also *faba crassa,* fat bean, but without any reason
assigned. It has been of late years transferred in England
to plants of the Goosefoot tribe, and more particularly to
the Good Henry, which a correspondent of Seeman's
Journal (vol. i. p. 151), asserts to have been used formerly

for fattening poultry ; a statement which requires confirmation, as there is no other English or foreign writer who mentions any such use of the Goosefoots, or the orpine.

Chenopodium Bonus Henricus and Atriplex patula, L.

FEABE, FAPE, FABE, THAPE, THEABE, DE-, FAE-, FEA-, or FEAP-BERRY, different forms of an East Anglian very obscure name of the gooseberry. An interchange of an initial *f* and *th* is not uncommon. We find it, for instance, in the verb *fly*, Go. *pliuhan*, O.H.G. *fliohan*, G. *fliehen* ; and in *thatches*, a dialectic pronunciation of *vatches*, i.e. *fitches* or *vetches*, to be heard in some parts of Somerset. The name seems to have attached itself to the gooseberry through one of those blunders that have arisen from bad pictures. The melon, G. *pfebe*, L. *pepo*, is so figured in Tabernæmontanus (vol. ii. p. 184), as to look exactly like a gooseberry, and headed *Pfebe*, and from this, or an equally bad picture, the name may have been adopted. Loudon (in Arbor. Brit. ii. p. 972), considers it to be meant for *fever-berry*, a very improbable explanation. Wright would derive it from A. S. *þefe-þorn*. The use of the term seems to be confined at present to the eastern counties, where the green unripe fruit is called *Thape*, as in a word well known to Norfolk schoolboys, *Thape-pie*.

Ribes Grossularia, L.

FEATHER-FEW, FEDDER-FEW, or FEATHER-FULLY, in Pr. Pm. FEDER-FOY, the feverfew, from confusion of name with the feather-foil, Pyrethrum Parthenium, L.

FEATHER-FOIL, feathery leaf, from *feather* and *foil*, L. *folium*, a name descriptive of its finely divided leaves,

Hottonia palustris, L.

FEATHER-GRASS, from its feathery awn,

Stipa pennata, L.

FELWORT, the gentian, a name that looks like an anomalous compound of L. *fel*, gall, and *wort*, but apparently is

a modified form of A.S. *feld-wyrt*, which there is little
doubt arose from a confusion between L. *fel* and A. S. *feld*,
Gentiana, L.

FELON-WORT, or -WOOD, the bittersweet, from its use in
curing whitlows called in Latin *furunculi*, little thieves,
that is, felons,　　　　　　　　Solanum Dulcamara, L.

FEN-BERRY, from its growing in fens, the cranberry,
Vaccinium Oxycoccos, L.

FEN-RUE, from its divided rue-like leaves and place of
growth,　　　　　　　　Thalictrum flavum, L.

FENKEL, and FENNEL, M.Lat. *fanculum*, from L. *fœniculum*, F. *fenouil*,　　　　　　　　F. vulgare, Gärt.

　,,　Dog's-, from some similarity of its leaf to fennel,
and its bad smell,　　　　　　　　Anthemis Cotula, L.

FENNEL-FLOWER, from its fennel-like finely divided
leaves,　　　　　　　　Nigella damascena, L.

FERN, A.S. *fearn*, G. *farnkraut*, Du. *varenkruidt*, a word
of obscure origin, but from the appended *kraut* and *kruidt*
to the G. and Du. synonyms, one that seems to be expressive of some use or quality. It has been, with great
plausibility, referred to forms of the word *feather*, G. *feder*,
Sl. *pero*, Gr. πτερις and πτερον, and suggested that it may
be connected with Skr. *parna*, which means both a leaf and
a feather, and with L. *frons*. It seems a more easy and
natural explanation of the word to trace it to the use of
these plants for littering cattle, A.S. *fear*, G. *farr*, Du.
var, a bullock, in which the change of the letters exactly
corresponds to that which takes place in the names of *fern*
in the same languages. It might also be connected with
the name of the pig, A.S. *fearh*, Du. *varken*, G. *ferkel* and
ferken, L. *verres*, Skr. *varâha*. Matthioli (l. iv. c. 179),
speaking of Pteris, and Filix mas, says: " Utriusque
radice sues pinguescunt ;" and in a charter of A.D. 855,
in Thorpe's Diplomatarium, p. 113, *pascua porcorum* is
translated *fearnleswe*. In some old German works the

word is *varm*, and as the Scandinavian name is *orm-gräs*, worm- or snake-grass, from the involuted vernation of the frond, it may be worth consideration, whether these words, *varm* and *orm*, may not be the Lat. *vermis*, and *farn* a corruption of it.   But, as J. Grimm says, "tiefes dunkel ruht auf der wurzel."   Filix.

FERN, BLADDER-, from its small vesicular spore-cases,
                                        Cystopteris, Ber.

„   BRISTLE-, from the bristle that projects beyond its receptacle,   Trichomanes, Sm.

„   FEMALE-, of old writers, not the species now called *Lady-fern*, but the brake, from Gr. θηλυπτερις, a name that is now assigned to another one different from both of these.   That of the herbals is   Pteris aquilina, L.

„   FILMY-, from its transparent filmy texture,
                                        Hymenophyllum, L.

„   FINGER-,   Ceterach officinarum, W.

„   FLOWERING-, from its conspicuous spikes of fructification,   Osmunda regalis, L.

„   HARD-, from the rigid texture of the frond,
                                        Blechnum boreale, Sm.

„   HOLLY-, from its prickly fronds,
                                        Aspidium Lonchitis, Sw.

„   LADY-, a mere translation of *filix foemina*, a former Latin name of the brake, Gr. θηλυπτερις, which was given to it from some caprice, and without reference to sex as now understood, and has been transferred to the species now called so, for no better reason than the delicacy of its foliage,   Asplenium Filix foemina, Bern.

„   MAIDEN-HAIR-,   Adiantum Capillus Veneris, L.

„   MALE-, a translation of its old Latin name,
                                        Aspidium Filix mas, Sw.

„   MARSH-,   Aspidium Thelypteris, Sw.

„   MOUNTAIN-,   A. Oreopteris, Sw.

„   OAK-, of modern botanists,
                                        Polypodium Dryopteris, L.

FERN, OAK-, of the herbalists, with a belief that such plants of it as grew upon the roots of the oak-tree, were of greater medicinal power: "quod nascit super radices quercus, est efficacius :" (Herbarius, c. 103.) the common polypody,   Polypodium vulgare, L.

   ,,   PARSLEY-,   Allosorus crispus, Bern.

   ,,   SCALY-, from the scales on the frond,
             Ceterach officinarum, W.

   ,,   SHIELD-, from the coverings of its spore-cases,
             Aspidium, Sw.

   ,,   WALL-, from its ordinary place of growth,
             Polypodium vulgare, L.

FESCUE, from the L. *festuca*, by change of *t* to *c*,
             F. ovina, L. etc.

FEVERFEW, L, *febrifuga*, from its supposed febrifuge qualities,   Pyrethrum Parthenium, L.

FIELD CYPRESS, the ground pine, from its terebinthinate odour and divided leaves,   Ajuga Chamæpitys, Schr.

FIELD MADDER,   Sherardia arvensis, L.

FIG, Fr. *figue*, L. *ficus*,   F. Carica, L.

FIG-WORT, from its use, on the doctrine of signatures, in the disease called *ficus*,   Scrophularia, L.
and also, for the same reason,   Ranunculus Ficaria, L.

FILBERT, formerly spelt *Filberd*, and *Fylberde*, said by some to have been so called after a king Philibert; by Wedgwood explained as *Fill-beard;* but more probably a barbarous compound of *phyllon* or *feuille*, leaf, and *beard*, to denote its distinguishing peculiarity, the leafy involucre projecting beyond the nut,   Corylus Avellana, L.

FINCKLE, G. *fenchel*, Du. *venkel*, from L. *fœniculum*, fennel,   F. vulgare, L.

FINGER-FLOWER, G. *finger-hut*, L. *digitalis*, from the resemblance of its flower to the finger of a glove,
             D. purpurea, L.

FIORIN, Erse *fearh*, grass,   Agrostis alba, L.

FIR, O.H.G. *furaha*, Dan. *fyrr*, Sw. *furu*, the *fire*-tree, the most inflammable of woods. Etymologically the word is identical with L. *quercus*, the initial *qu* having, as in *quinque*, an *f* for its representative in the Germanic languages. See Max Müller, Lectures, 2nd course, p. 224.
Pinus and Abies.

,, SCOTCH-, from its being found indigenous upon the mountains of Scotland,   P. sylvestris, L.

,, SILVER-, from its white trunk,   P. Picea, L.

,, SPRUCE-, from G. *sprossen*, sprout, see SPRUCE,
Abies excelsa, Poir.

FIR-MOSS, a mossy looking plant like a little fir-tree,
Lycopodium Selago, L.

FIST-BALLS, A.S. *fist*, G. *feist*, Du. *veest*, crepitus, Norw. *fissopp* and *fisball*. See PUCKFIST.   Lycoperdon, L.

FITCH, an old spelling of Vetch, from L. *Vicia*.

FIVE-FINGER-GRASS, or FIVE-LEAF, Sw. *finger-ört*, Gr. πενταδακτυλον, five-fingered, ancient Gallic as quoted by Dioscorides, (iv. c. 42), πεμπεδουλα; by Apuleius, (c. 2,) *pompedulon*; a word meaning five-leaf, a plant so called from its five leaflets,   Potentilla reptans, Sib.

FLAG, from its petals hanging out like banners, properly the species of   Iris, L.

,, CORN-,   Gladiolus, L.

,, SWEET-,   Acorus Calamus, L.

,, YELLOW-,   Iris Pseudacorus, L.

FLAMY, in Mrs. Kent's Flora Domestica given as a name of the pansy, and explained " because its colours are seen in the flame of wood." It is the translation of Lat. *Viola flammea*.   V. tricolor, L.

FLAW FLOWER, from *flaw*, a gust of wind, a translation of Gr. ἀνεμωνη, from ἀνεμος, wind. See ANEMONY.
Anemone Pulsatilla, L.

FLAX, G. *flachs*, Du. *vlas*, Fr. *filasse*, M. Lat. *filassium*, yarn, from *filare*, spin, L. *filum*, a thread,
Linum usitatissimum, L.

FLAX, DWARF- or PURGING- or FAIRY-,

                                      L. catharticum, L.

   ,, TOAD-,   Linaria vulgaris, L.

FLAX-SEED, from the resemblance of its seed-pods to flax bolls,   Radiola Millegrana, L.

FLAX-WEED, from its leaves resembling those of flax,

                              · Linaria vulgaris, L.

FLEA-BANE, from its supposed power of destroying fleas,

                               Inula Pulicaria, L.

   ,, BLUE-,   Erigeron acre, L.

FLEA-WORT, from its keeping off fleas,

                              Inula Conyza, DC.

FLEUR DE LIS, see FLOWER DE LUCE.

FLIX- or FLUX- WEED, from its use in dysentery, a disease that was formerly called *flix*,

                        Sisymbrium Sophia, L.

FLOAT- or more properly FLOTE-GRASS, not so much from its floating on the surface of the water, as from its abounding in *floted*, or irrigated meadows,

                         Poa fluitans, Scop.

and also in some works

   Alopecurus geniculatus, and Catabrosa aquatica, L.

FLORIMER, or FLORAMOR, Fr. *fleur d'amour*, from a mis-understanding of its Latin name, *Amaranthus*, as though a compound of *amor*, love, and *anthus*, flower,

                             A. tricolor, L.

FLOWER DE LUCE, Fr. *fleur de Louis*, from its having been assumed as his device by Louis VII. of France: "Ce fut Louis VII. dit le Jeune, A.D. 1137, qui chargea l'écu de France de fleurs de lis sans nombre," Montf. The flower that he chose seems to have been a white one; for Chaucer says:

      His nekke was white as is the *flour de lis*:

and there is a legend that a shield charged with these flowers was brought to Clovis from heaven while engaged in a battle against the Saracens (H. Pyne, England and France, in 15th cent. p. 23). It had already been used by the other French kings, and by the Emperors of Constantinople; but it is a question what it was intended in the first place to represent. Some say a flower, some a halbert-head, some a toad. See Notes and Queries, 29 Mar. 1856. *Fleur de Louis* has been changed to *Fleur de Luce*, *Fleur de lys*, and *Fleur de lis*.   Iris, L.

FLOWER OF BRISTOW, or -OF CONSTANTINOPLE, the scarlet lychnis, the latter name from its growing wild near the Turkish capital, the former from its colour being "Bristol red," as in the expression:

"Her kirtle Bristol red."

See Chambers' Book of Days, i. p. 801.

Lychnis chalcedonica, L.

FLOWER GENTLE, the floramor, Fr. in Cotgrave *la noble fleur*, from its resemblance to the plumes worn by people of rank,   Amaranthus tricolor, L.

FLOWERING FERN, from its handsome spikes of fructification,   Osmunda regalis, L.

FLOWERING RUSH, L. *juncus floridus*, a plant with a rush-like stem, and growing in the water, with a fine head of flowers, called by Lobel *Juncus cyperoides floridus*, "*Juncus*," saith he, "for that his stalke is like the rush; *cyperoides*, because his leaves do resemble Cyperus; *floridus*, because it hath on the top of every rushie stalke a fine umbel or tuft of small flowers in fashion of the Lilie of Alexandria." Gerarde, p. 27.

Butomus umbellatus, L.

FLOWK-, or FLOOK-WORT, from its being supposed to give sheep the disease of the liver, in which parasites resembling the *flook-* or *flounder*-fish are found,

Hydrocotyle vulgaris, L.

FLUELLIN, Du. *fluweelen*, downy, velvety, Fr. *velvote*, and not, as Parkinson states, a Welsh word:

„ MALE-, of Gerarde, Hill, Curtis, and others, from its soft velvety leaves,   Linaria spuria, L.

„ FEMALE-,   Veronica Chamædrys, L.

FLYBANE, from its being used mixed with milk to kill flies,   Agaricus muscarius, L.

FLY HONEYSUCKLE, from confusion with an Apocynum that catches flies by the proboscis under its anthers, the A. androsæmifolium, L. and whose flowers are somewhat similar to those of the upright honeysuckle,

Lonicera Xylosteum, L.

FLY ORCHIS, from the resemblance of its flower to a fly,

Ophrys muscifera, Hud.

FOLEFOOT, from the shape of its leaf,

Asarum europæum, L. and Tussilago Farfara, L.

FOOL'S PARSLEY, from its being a poisonous plant, which only fools could mistake for parsley,

Æthusa Cynapium, L.

FOREBITTEN MORE, bitten-off root, see DEVIL'S-BIT, *more* or *mor* having formerly had the sense of "root," as it has still in the Western counties,   Scabiosa succisa, L.

FORGET-ME-NOT, a name that for about forty years has been assigned to a well known blue flower, a Myosotis, but which for more than 200 years had in this country, France, and the Netherlands, been given to a very different plant, the ground-pine, Ajuga Chamæpitys, on account, as was said, of the nauseous taste that it leaves in the mouth. It is to this plant exclusively that we find it assigned by Lyte, Lobel, Gerarde, Parkinson, and all our herbalists from the middle of the fifteenth century, and by all other botanical authors who mention the plant, inclusive of Gray in his Natural Arrangement published in 1821, until it was transferred with the pretty story of a drowned lover to that which now bears it. This had always been called in England Mouse-ear Scorpion-grass. In Germany, Fuchs in his Hist. Plant. Basil, 1542, gives the name *Vergiss nit mein* to the Teucrium Botrys, L. under the Lat. synonym of Chamædrys vera fœmina. His excellent plate at p. 870 leaves no doubt as to the species he meant. In Denmark a corresponding name, *Forglemm mig icke*, was given to the Veronica chamædrys. At the same time it would seem that in some parts of Germany the Myosotis palustris was known as the Echium amoris, and *Vergiss mein nicht*, as at the present day; and in Swedish the Echium aquaticum, the same plant, was called *förgät mig icke*. Some idea of the confusion will be seen in Mentzel's Index Nominum Plantarum, Berlin, 1682. Cordus on Dioscorides, in 1549, and Lonicerus assign it to Gnaphalium leontopodium, L. while the Ortus Sanitatis (Ed. 1536, ch. 199), and Macer de virtutibus herbarum (Ed. 1559), like the Danish Herbalists, give it to the Veronica Chamædrys, L. This latter seems to be the plant to which the name rightfully belongs, and to which it was given in reference to the blossoms falling off and flying away. See SPEEDWELL. From this plant it will have been transferred to the ground-pine through a confusion in respect to which species should properly be called *Chamædrys*, and as both these very different plants were taken for the Chamædrys of Pliny, the popular name of the one passed to the other. Two circumstances about it are curious; first, how the name could be transferred from the ground-pine to the scorpion-grass without the change being noticed by a single author of all our floras, general and local; and secondly, how easily a good story is got up, and widely spread about the world, to match a name. The blossoms fall from a Veronica, and it is called "Speedwell! and "Forget-me-not." The name passes to a plant of nauseous taste, the ground-pine, and Dalechamp explains it as expressive of this disagreeable quality. It attaches itself to a river-side plant, and the story books are ready with a legend. We learn from Mills's History of Chivalry that a flower that bore the name of "Soveigne vous de moy," was in the fourteenth century woven into collars, and worn by knights, and that one of these was the subject of a famous joust fought in 1465 between the two most accomplished knights of England and France. What the flower was, that was so called, it would be only possible to ascertain by inspection of one of these collars. The German name *Ehrenpreis*, prize of honour, which has always been given to the speedwell, almost proves that this was the one. There is certainly no ground for assuming that it was the same as our present "Forget-me-not." The story of this latter, in connexion with the two lovers, will be found in Mills's work, vol. i. p. 314, and it now bears a name corresponding to our own in nearly every European language: as Fr. *Ne m'oubliez pas*, G. *Vergiss mein nicht*, Da. *Kiærminde*, Sw. *Forgät mig icke*, etc., and is worked into numberless rings and other ornaments.   Myosotis palustris, L.

FOUR-LEAVED GRASS, a plant with four leaves only, the Herb Trulove,   Paris quadrifolia, L.

FOXGLOVE, a name that is so inappropriate to the plant, that many explanations of it have been attempted, by which it might appear to mean something different from the glove of a fox. Its Norwegian names, *Rev-bielde*, foxbell, and *Reveleika*, fox music, are the only foreign ones that allude to that animal; and they explain our own, as having been in the first place *foxes-glew*, or music, A.S. *gliew*, in reference to a favourite instrument of earlier times, a ring of bells hung on an arched support, a tintinnabulum, which this plant, with its hanging bell-shaped flowers, so exactly represents. Its present Latin name, *Digitalis*, was given to it by Fuchs with the remark that up to that time, 1542, there was none for it in Greek or Latin.   D. purpurea, L.

FOX-TAIL-GRASS, from the shape of the spike,
                    Alopecurus pratensis, L.

„ Slender-,   Alopecurus agrestis, L.

Framboise, a French corruption of Du. *brambezie*, bramble-berry, the raspberry,   Rubus idæus, L.

Franke, from "the property it hath to fatten cattle," as Lyte tells us, ch. 38; *franke* meaning a stye or stall, in which cattle were shut up to be fattened,
                    Spergula arvensis, L.

French bean, a foreign bean, *French* being used to express what in German would be called *wälsch*, anything from an outlandish country,   Phaseolus vulgaris, L.

French cowslip,   Primula Auricula, L.

French Grass, sainfoin, L. *fœnum Burgundiacum*,
                    Onobrychis sativa, L.

French Honeysuckle, from the resemblance of its flowers to large heads of honeysuckle clover,
                    Hedysarum coronarium, L.

French Lavender,   Lavandula Stœchas, L.

French Nut, the walnut,   Juglans regia, L.

French Sorrel, the wood-sorrel, Oxalis acetosella, L.

French Sparrow-grass, the name under which are sold in the Bath market to be eaten as asparagus, the sprouts of the spiked Star of Bethlehem,
                    Ornithogalum pyrenaicum, L.

French Wheat, the buckwheat,
                    Polygonum Fagopyrum, L.

French Willow, from its leaves somewhat resembling those of the willow,   Epilobium angustifolium, L.

Fresh-water Soldier, from its sword-shaped leaves,
                    Stratiotes aloides, L.

Friar's cap, from its upper sepals resembling a friar's cowl, the wolfsbane,   Aconitum Napellus, L.

„ crown,   Carduus eriophorus, L.

Fritillary, M.Lat. *fritillaria*, sc. tabula, a checkerboard, from *fritillus*, a dicebox, on account of its checkered petals,   F. Meleagris, L.

FROG-BIT, L. *morsus ranæ*, from an idea that frogs ate it,
Hydrocharis Morsus ranæ, L.

FROG-FOOT, a name that in the Stockholm Med. M.S.
l. 783. is with more reason assigned to the vervain, the
leaf of which, in its general outline, somewhat resembles
the foot of this animal:

> *Frossis fot* men call it,
> For his levys are lyke the frossys fet.

In modern works it is transferred to the duckmeat,
Lemna, L.

FROG-GRASS, from its growing in mire,
Salicornia herbacea, L.

FROG'S-LETTUCE,   Potamogeton densus, L.

FROST-BLITE, a blite whitened as by hoar-frost,
Chenopodium album, L.

FULLER'S HERB, L. *herba fullonum*, from its taking out
stains from cloth, a purpose for which it is said by Tragus,
c. 131, to have been used by the monks,
Saponaria officinalis, L.

FULLER'S THISTLE, the teasel, Dipsacus fullonum, L.

FUMITORY, Fr. *fume-terre*, L. *fumus terræ*, earth-smoke,
from the belief that it was produced without seed from
vapours rising from the earth. The words of Platearius,
a great authority in his day, are: " Dicitur *fumus terræ*,
quod generatur a quadam fumositate grossa, a terra reso-
luta, et circa superficiem terræ adherente." See also the
Ortus Sanitatis, Mayence, 1485, ch. 176, and the Grete
Herball, cap. clxxi. And this extraordinary account of it
is given not only by the ignorant authors of the Ortus
Sanitatis and the Grete Herball, and the writers in the
dark ages from whom they copied, but is repeated by
Dodoens, and other learned men, his cotemporaries. Pliny
merely says, (l. xxv. c. 13), that it took its name from
causing the eyes to water when applied to them, as smoke

does: "Claritatem facit inunctis oculis delachrymationem-
que ceu fumus, unde nomen accepit καπνος."

Fumaria officinalis, L.

FURZE, sometimes spelt FURRES, A.S. *fyrs*, a name of
obscure derivation, as are those of so many of our commonest plants; apparently from *fir*, these bushes being,
like the coniferous trees, a common firewood or fuel; but
perhaps from Fr. *forest*, as though that word meant a place
of firs, as hyrst, carst, hulst, gorst, etc., the places or
thickets of erica, carices, ulex and gorra, from M. Lat.
words in *cetum*;   Ulex europæus, L.

  ,, NEEDLE-, from its finely pointed slender spines,

Genista anglica, Hud.

FUSS-BALLS, Fr. *vesse*,   Lycoperdon, L.

GADRISE, and GAITRE, see GATTER.

GALE, or SWEET GALE, in Turner's herbal GALL, and in
Somersetshire, he tells us, GOUL and GOLLE, in Pr. Pm.
gawl, *gavl*, or *gawyl*, A.S. and Du. *gagel*, corruptions apparently of *galangale*, a name that it may have acquired from
its fragrance while burning, and which, through its intense
bitterness, has become confounded with *gall*,

Myrica Gale, L.

GALANGALE, It. and Sp. *galanga*, O. Sp. *garingal*, G.
galgant, from *chalan*, spice, a Persian word transferred
to a marsh plant, the roots of which are valued for their
aromatic quality,   Cyperus longus, L.

GALLOW-GRASS, Ger p. 572, a cant name for hemp, as
furnishing halters for the gibbet,   Cannabis sativa, L.

GANG-FLOWER, *flos ambarvalis*, the milkwort, from its
blossoming in *Gang*-week, A.S. *gang-dagum*, three days
before the Ascension, when processions were made in imitation of the ancient Ambarvalia, to perambulate the
parishes with the Holy Cross and Litanies, to mark their
boundaries, and invoke the blessing of God upon the

crops ; on which occasions, says Bishop Kennett, "the maids made garlands of it and used them in those solemn processions." So also Gerarde, 1st ed. p. 450. It was for the same reason called Cross-, Rogation-, and Procession-flower.   Polygala vulgaris, L.

GARAVANCE, the chick-pea, or gram, Sp. *garavanzo*, Bask. *garau*, corn, and *anzua*, dry (Diez),
                                        Cicer arietinum, L.

GARLICK, A.S. *gar*, a spear, and *leac*, plant, from its tapering acute leaves ; or from the nutritive and stimulant qualities ascribed to it by the ancient northern poets as being the " war plant,"   Allium sativum, L.

GARLICK-WORT, from its smell,
                                    Erysimum Alliaria, L.

GARNET-BERRY, the red currant, from its rich red colour and transparency,   Ribes rubrum, L.

GATTER, GATTEN, GADRISE, or GATTERIDGE, names of several hedgerow trees and shrubs, as the spindle, the cornel, and wild Guelder-rose,   Evonymus europæus,
          Cornus sanguinea, and Viburnum Opulus, L. derived, respectively, *Gatter*, in Chaucer *Gaitre*, from A.S. *gad*, a goad, and *ter* i.e. *treow*, tree ; *Gatten*, from *gad*, and *tan*, twig ; *Gadrise*, from *gad* and *hris*, a rod, Da. and Du. *riis*, a shrub, and *Gatteridge*, Fr. *verge sanguine*, from *gaitre rouge*, in reference to the red colour of the twigs and autumnal foliage of the spindle and cornel tree.   *Gad* is still used in our Western counties for a picked stick in the term *spar-gad*, a stick pointed at both ends to spar or fasten down thatch.

GAZLES, in Sussex and Kent, the black currant, apparently corrupted from Fr. *groseilles*,
                                    Ribes nigrum, L.

GEAN, the wild cherry, Fr. *guisne*, Pol. *wisn*, Boh. *wissne*, in European Turkey *wischna*, Wal. *visini*, M.Gr. βισινος, the two last words being identical with the Slavonian, as far as they can be written with Greek letters. We may conclude from this identity, and from the great quantity of pipe-sticks of it exported every year from Turkey, that the name originated in that country. The Dalmatians will have Italianized *wischna* into *viscina*, and under this name it will have been conveyed to Italy, and thence into France, where by the usual process of changing *v* or *w* to *gu*, and dropping the sound of *s* and *sc* before *n*, *viscina* became *guisne*, and crossed into England as *gean*. But the Italians will have regarded *riscina* as a diminutive in *ina* from *viscia*, and have replaced it, from some motive of euphony, by *visciola*, its present name, as they have formed *bisciuola* from *biscia*, *pesciuolo* from *pesce*, etc., and hence the German *weichsel*, which will not only represent the same tree, but the same word, as our *Gean*. It is to be observed that the " Bird cherry " is not this species, although this is the one called so by botanists in Latin systematic works. Prunus avium, L.

GENTIAN, from some Illyrian king named Gentius,
Gentiana, L.

GERANIUM, Gr. γερανιον, from γερανος, crane, the cranesbill, from the shape of the seed vessel, a genus that once included the Pelargonia, which in popular language are still called so, G. molle, pratense, etc.

GERARD, see HERB GERARD.

GERMAN MADWORT, Du. *meed*, madder, and *wort*, root, so called from the red dye yielded by its roots, and its being used in Germany, Asperugo procumbens, L.

GERMANDER, Fr. *gamandrée*, from L. *chamædrys*, by insertion of an *n* before *d* for euphony, Gr. χαμαι, ground, and δρυς, oak, so named from the fancied likeness of its leaves to those of an oak, a name usually given to the
Teucrium Chamædrys, L.
,, WATER-, Teucrium Scordium, L.
,, WOOD-, Teucrium Scorodonia, L.

GERMANDER CHICKWEED, the male Chamædrys of the herbalists,   Veronica agrestis, L.

GILL, GILL-GO-BY-GROUND, GILL-CREEP-BY-THE-GROUND, GILL-RUN-BITH-GROUND, the ground-ivy, from its name *Gill*, that was given to it from its being used in fermenting beer, Fr. *guiller*, a word still retained in the eastern counties, getting mixed up with another meaning of *Gill*, that of a young woman, a girl; the *go-by-ground*, etc. referring to the creeping habit of the plant. See HAY-MAIDS.   Nepeta Glechoma, Benth.

GILLIFLOWER, formerly spelt *gyllofer* and *gilofre*, with the *o* long, from Fr. *giroflée*, It. *garofalo*, in Douglas's Virgil *jereflouris*, words formed from M. Lat. *garoffolum*, *gariofilum*, or, as in Albert Magn. (l. vi. c. 22), *gariofilus*, corrupted from L. *caryophyllum*, a clove, Gr. καρυοφυλλον, and referring to the spicy odour of the flower, which seems to have been used in flavouring wines to replace the more costly clove of India. The name was originally given in Italy to plants of the Pink tribe, especially the carnation, but has in England been transferred of late years to several cruciferous plants. That of Chaucer and Spenser and Shakspeare was, as in Italy,   Dianthus Caryophyllus, L. that of later writers and gardeners,

Matthiola and Cheiranthus, L.

Much of the confusion in the names of plants has arisen from the vague use of the French terms *Giroflée*, *Oeillet*, and *Violette*, which were, all three of them, applied to flowers of the Pink tribe, but subsequently extended, and finally restricted in English to very different plants. *Giroflée* has become *Gilliflower*, and passed over to the Cruciferæ, *Oeillet* been restricted to the Sweet Williams, and *Violette* been appropriated to one of the numerous claimants of its name, the genus to which the pansy belongs.

   „ CLOVE-,   Dianthus Caryophyllus, L.

,, MARSH-, the ragged-Robin,

Lychnis flos cuculi, L.

,, QUEEN'S-, or ROGUE'S-, or WINTER-, the Dame's
violet, Hesperis matronalis, L.

,, STOCK-, Matthiola incana, L.

,, WALL-, of old books, Cheiranthus Cheiri, L.

,, WATER-, of Lyte's Herball,

Hottonia palustris, L.

GILLIFLOWER-GRASS, in Aubrey's Wilts, p. 49. See
CARNATION-GRASS.

GIPSEY-WORT, so called, says Lyte, " bycause the rogues
and runagates which call themselves Egyptians, do colour
themselves black with this herbe,"

Lycopus europæus, L.

GITH, L. *gith*, a name now applied to the corn-cockle,

Agrostemma Githago, L.

GLADDON, GLADEN, GLADER, GLADWYN, names of the
stinking iris usually derived from L. *gladiolus*, a small
sword, in allusion to its sword-shaped leaves, but which
have more probably arisen from confusion of its Dutch
name *lisch*, with O.French *leesche*, gladness. If from
gladiolus, they will be plurals of *glad*; but as in herbal
nomenclature the plant is called *spatula*, a tool used in
smoothing, they may be related to Du. *glad*, smooth.

Iris fœtidissima, L.

GLADIOLE, L. *gladiolus*, a small sword,

,, WATER-, the flowering rush,

Butomus umbellatus, L.

GLASSWORT, from furnishing ashes for glass-making,

Salicornia herbacea, L.

,, PRICKLY-, Salsola Kali, L.

GLASTONBURY THORN, a variety of whitethorn, so called
from the place where it was first cultivated.

GLOBE FLOWER, from its globular form,

Trollius europæus, L.

GLOBE THISTLE, from its globular inflorescence,

                                Echinops, L.

GOAT'S BEARD, from its long coarse pappus, a translation of its Gr. name τραγοπωγων,

                       Tragopogon pratensis, L.

GOAT-WEED, from its Greek name, αιγοποδιον,

                       Ægopodium Podagraria, L.

GOLD-APPLES, Fr. *pommes d'or*, from their colour before maturity, tomatoes,   Solanum Lycopersicum, L.

GOLD OF PLEASURE, a name which Gerarde and Parkinson attempt to explain by telling us that "the poore peasant doth use the oile in banquets, and the rich in their lampes." This seems to be a way of getting over a difficulty by forcing a sense upon it. We learn from Gerarde that an oil was imported from Spain as "Oleo de Alegria," this latter word Alegria, being the name of another oil-plant, a sesamum, and it would seem that this "Oleo de Alegria" has become corrupted to "Oro de alegria," gold of pleasure, and applied to a very different species, the source of a spurious oil, passed off upon the public for the Spanish. Whether *Alegria* was applied to the sesamum in the sense of "pleasure," or is an Arabic word beginning with *al*, it is irrelevant to enquire.   Camelina sativa, L.

GOLD-CUPS, from A.S. *copp*, a head, a button, or stud, and like King-cup, Gilt-cup, and Butter-cup, representing the Fr. *bouton d'or*, the bachelor's button,   Ranunculus, L.

GOLD-KNOBS, -KNAPPES, or -KNOPPES, A.S. *cnæp*, a button, Du. *knoop*. See GOLD-CUP.

GOLDE, in our old poets the marigold, supposed from its yellow flowers, to have been the χρυσανθεμον, or gold flower of the Greeks,   Calendula officinalis L.

GOLDEN-CHAIN, from its long racemes of yellow flowers, Du. *goude keten*, in Sweden more tastefully called *guldregn*, golden rain,   Cytisus Laburnum, L.

GOLDEN-ROD, Lat. *virga aurea*, from its tall straight stalk of yellow flowers, Solidago Virga aurea, L.

GOLDEN SAMPHIRE, from its thick samphire-like stems, and its golden flowers, Inula crithmoides, L.

GOLDEN SAXIFRAGE, from its yellow flowers,
Chrysosplenium, L.

GOLDILOCKS, Gr. χρυσοκομη, from χρυσος, gold, and κομη, hair, Chrysocoma Linosyris, L.
also Ranunculus Auricomus, L.

GOLDINS, or GOLDINGS, Du. *gulden*, golden, a florin, from the yellow colour and flat round shape of its flowers, the source of the numerous Scotch names applied to the marigold, the marsh marigold, and other yellow flowers, such as *Gowlan, Gowan, Gool, Goule*, etc. See below GOOLS. By *Goldin* is usually meant the corn-marigold,
Chrysanthemum segetum, L.

GOOD HENRY, or GOOD KING HARRY, G. *guter Heinrich*, Du. *goeden Henrik*, an obscure name, which Dodoens tells us (p. 651) was given to the plant to distinguish it from another, a poisonous one, called *Malus. Henricus;* but why they were either of them called *Henricus*, we are not told. Cotgrave gives the name *Bon Henry* to the Roman sorrel, Rumex scutatus, L. as well as to the allgood, the plant to which it is usually assigned. Cordus on Dioscorides, Frankf. 1549, calls it " *Weyss heyderich*, vel ut alii volunt, *Gùt heynrich*." It has nothing to do with our Harry the 8th, and his sore legs, to which some have thought that it referred. Chenopodium Bonus Henricus, L.

GOOLS, GULES, GOWLES, GUILDES, GOULANS, GOWANS, and GOLDS. See under GOLDINS.
Calendula officinalis, Caltha palustris, and Chrysanthemum segetum, L.

GOOSE AND GOSLINGS, or GANDERGOSSES, from the flowers being shaped like little goslings,
Orchis Morio, and bifolia, L.

GOOSEBERRY, from the Fl. *kroes* or *kruys bezie*, Sw. *krusbär*, a word that bears the two meanings of "cross-" and "frizzle-berry," but was given to this fruit with the first meaning in reference to its triple spine which not unfrequently presents the form of a cross. This equivocal word was misunderstood and taken in its other sense of "frizzle-berry," and translated into German and herbalist Latin *kraüsel-beere* and *uva crispa*. Matthioli (ed. Camerarü, 1586) gives its German synonym correctly, as *kreuzbeer*. Lobel also (Krydtb. pt. ii. p. 239) gives it as Flem. *kroesbesien*, G. *kruzbeer*. The Fr. *groseille*, and Span. *grosella* are corruptions of G. *kraüsel*.

Ribes Grossularia, L.

GOOSEBILL, or GOOSESHARE, clivers, from the sharp serrated leaves being like the rough-edged mandibles of a goose,   Galium Aparine, L.

GOOSECORN, from its growth on commons where geese are commonly reared, and the grain-like appearance of the capsules,   Juncus squarrosus, L.

GOOSE-FOOT, from the shape of its leaf,

Chenopodium, L.

GOOSE-GRASS, in Ray by mistake GOOSE-GREASE, Pr. Pm. *gosys gres*, clivers, from a belief that goslings feed on it, (R. Turner, Bot. p. 71), and that "geese help their diseases with it," (Lupton, No. 60).

Potentilla anserina, L.

GOOSE-HEIRIFFE of W. Coles's Adam in Eden; A.S. *gos*, a goose, and *hegerife*, hedge-reeve, from its attaching itself to geese, while they pass through a hedge. The occurrence of the name in this work of W. Coles is singular as an instance of the retention into the seventeenth century of an Anglo-Saxon word no longer understood. The name is still retained in some counties as *hariff*. The *Gooshareth*, *Goshareth*, and *Gooseshareth* of W. Turner's Herball seem to be corruptions of *Goosehariff* with a change of *f* to *th* that is not uncommon in provincialisms, as for instance in the case of *fape* and *thape*. Galium Aparine, L.

GOOSE-TANSY, a plant with tansy-like leaves, which Ray says is called so "because eaten by geese;" but perhaps like crow's garlick, swine's cress, and dog's mercury, the name may imply merely a tansy for a goose, a spurious tansy. Potentilla anserina, L.

GOOSE-TONGUE, from its finely serrated leaves,

Achillæa Ptarmica, L.

GORSE, A.S. *gorst*, Wel. *gores* or *gorest*, a waste, M.Lat. *gorassi* or *gorra*, brushwood, used in Stat. Montis reg. p. 236 : "salicum, *gorrarum* et *gorassorum* non portantium fructus comestibiles." Ulex europæus, L.

GORY-DEW, from its resemblance to blood drops,

Palmella cruenta, Agh.

GO-TO-BED-AT-NOON, from its early closing, the goat's beard, Tragopogon pratensis, L.

GOURD, Fr. *gourde*, from *gougourde*, L. *cucurbita*,

C. Pepo, L.

GOUT IVY, M.Lat. *Iva arthritica*, from being, as Parkinson says, "powerful and effectual in all the pains and diseases of the joints, as gouts, cramps, palsies, sciatica, and aches," the ground pine, Ajuga Chamæpitys, L.

GOUT-WEED, or GOUT-WORT, from its supposed virtues in gout cases, Ægopodium Podagraria, L.

GOWAN, a north country word, usually derived from Gael. *gugan*, a bud, a flower, but clearly a corruption of *gowlan*, the Scotch form of *gulden*, as we see in the names of the troll-flower, which is called indifferently Lucken-*gowan* -*gollond*, or -*gowlan*, and Witch's *gowan*. Cf. *gawn*, a gallon measure, a milk-pail, in The Derby Ram. (Halliwell). In the glossaries it is usually explained as meaning merely "the daisy;" but appears in different parts of Scotland to be applied to the various buttercups, and the marsh-marigold, the dandelion, the hawkweeds, the corn-marigold, the globe flower, and indeed to almost any that is yellow. In the northern counties of England, according to Brockett, it is a yellow flower common in moist meadows, probably the marsh marigold.

GOWK-MEAT, from its blossoming when the cuckoo comes, the wood-sorrel,   Oxalis Acetosella, L.

GRAM, an Eastern name, the chick pea,

Cicer arietinum, L.

GRAPE HYACINTH, or GRAPE FLOWER, from its small round purplish flowers sitting in clusters on the stalk, like grapes,   Muscari racemosum, Mill.

GRASS, A.S. and Fris. *gœrs*, in nearly all other Germanic dialects *gras*, and radically connected with L. *gramen*. By the old herbalists *grass* is used in the sense of a herb generally, and often spelt *gres*, which has led to its being misspelt *grease* in several names. By botanists the term is confined to the order   Gramineæ.

GRASS, see under their specific names

,,	ARROW-	GRASS, MOUSE-TAIL-
,,	BALLOCK-	,, NIT-
,,	BENT-	,, OAT-
,,	BROME-	,, ORCHARD-
,,	CANARY-	,, PENNY-
,,	CARNATION-	,, PEPPER-
,,	CAT'S-TAIL-	,, PIGEON'S-
,,	COCK'S-FOOT-	,, PUDDING-
,,	CORD-	,, QUAKE-
,,	COTTON-	,, QUITCH-
,,	COUCH-	,, RAY-
,,	CRAB-	,, REED-
,,	CUCKOO-	,, RIB-
,,	DOG-	,, RIBBON-
,,	DOG'S-TAIL-	,, RIB- OR RYE-
,,	DOG'S-TOOTH-	,, SCORPION-
,,	DOOB-	,, SCURVY-

GRASS, FEATHER-	GRASS, SEA-
„ FESCUE-	„ SHAVE-
„ FIORIN-	„ SHELLY-
„ FINGER-	„ SHERE-
„ FIVE-FINGER-	„ SHORE-
„ FLOTE-	„ SPARROW-
„ FOXTAIL-	„ SPRING-
„ FRENCH-	„ SPURT-
„ GALLOW-	„ SQUIRREL-TAIL-
„ GRIP-	„ STANDER-
„ HAIR-	„ STAR-
„ HARD-	„ SWINE-
„ HARES-TAIL-	„ TIMOTHY-
„ HASSOCK-	„ TOAD-
„ KNOT-	„ TUSSAC-
„ LOB-	„ TWOPENNY-
„ LYME-	„ VERNAL-
„ MAT-	„ VIPER'S-
„ MEADOW-	„ WHITLOW-
„ MELICK-	„ WIRE-BENT-
„ MILLET-	„ WOOD-
„ MOOR-	„ WORM-

GRASS OF PARNASSUS, a plant described by the Greek writers as growing on that mountain,

Parnassia palustris, L.

GRASS-POLEY, from being considered by Cordus as a *pulegium* or *poley*, and having grassy leaves,

Lythrum hyssopifolium, L.

GRASS-VETCH, a vetch with grassy leaves,

Lathyrus Nissolia, L.

GRASS-WRACK, a wrack with long linear grass-like leaves,

Zostera marina, L.

GREEDS, A.S. *græd*, translated in Ælfric's glossary " ulva," a name of some water plant, now applied to the pondweed tribe,

Potamogeton, L.

GREEK VALERIAN, a plant mistaken for the Phu or Valerian of the Greek writers, Polemonium cæruleum, L.

GREEN-SAUCE, from its culinary use, Rumex acetosa, L.

GREEN-WEED, or GREENING-WEED, from its use to dye green, Genista tinctoria, L.

GREENS or GRAINES, in Lyte's Herball GRAYVES, Du. *Enden-gruen*, duck's herb, Lemna, L.

GRIGG, heath, Wel. *grúg*, related to W. *grwg* and *grig*, a rumbling noise. See BRAMBLE. Calluna vulgaris, Sal.

GRIMM THE COLLIER, the name of a humorous comedy popular in Q. Elizabeth's reign, "Grimm the collier of Croydon," given to the plant from its black smutty involucre, Hieracium aurantiacum, L.

GRIP-GRASS, from its gripping or seizing with its hooked prickles whatever comes in its way, Galium Aparine, L.

GROMELL, GRUMMEL, or GROMWELL, or GRAY MYLE, as Turner says it should be written, from *Granum solis* and *Milium solis* together. "That is al one," says the Grete Herball, "*granum solis* and *milium solis*." The apothecaries compromised the matter by combining them, as in the case of Asarabacca. Lithospermum officinale, L.

GROUND FURZE, Ononis arvensis, L.

GROUNDHEELE, G. *grundheil*, Fr. *herbe aux ladres*, so called from its having cured a king of France of a leprosy from which he had been suffering eight years, a disease called in German *grind*. Brunschwygk tells us (b. ii. ch. 5), that a shepherd had seen a stag, whose hind quarter was covered with a scabby eruption from the bite of a wolf, cure itself by eating of this plant and rolling itself upon it; and that thereupon he recommended it to his king.

Veronica officinalis, L.

GROUND NUT, or GRUNNUT, from its tuber having the flavour of a nut, Bunium flexuosum, W.

GROUND IVY, L. *hedera terrestris*, a name which at present is restricted to the Glechoma, but in the Stockholm

Med. M.S. 1. 864 (Archæol. v. xxx. p. 376) is given to the periwinkle :

> Parvenke is an erbe grene of colour;
> In tyme of may he beryth blo flour.
> His stalkys are so feynt and feye,
> That never more groweth he heye.
> On the grownde he rennyth and growe,
> As doth the erbe that hyth tunhowe.
> The lef is thicke, schinende, and styf,
> As is the grene ivy leef;
> Unche brod and nerhand rownde ;
> Men call it the *ivy of the grownde*.

From the periwinkle the name has been transferred to a labiate plant, Nepeta Glechoma, Benth.

GROUND PINE, Gr. χαμαιπιτυς from χαμαι, ground-, and πιτυς, pine, so called from its terebinthinate odour, the forget-me-not of all authors till the beginning of this century, Ajuga Chamæpitys, Sm.

GROUNDSEL, in a MS. of the fifteenth century *gronde-swyle*, A.S. *grundswelge*, ground glutton, from *grund*, ground, and *swelgan*, swallow, still called in Scotland and on the Eastern Border *grundy-swallow*, Senecio vulgaris, L.

GUELDER ROSE, from its rose-like balls of white flowers, and Gueldres, its native country, a variety of the water-elder, Viburnum Opulus, L.

GUERNSEY-LILY, from its occurrence on that island, Nerine sarniensis, W.

GULF-WEED, from its floating on the gulf-stream, Fucus natans, L.

GUINEA HEN, from its Latin name, *Meleagris*, given to it from its petals being spotted like this bird, a native of the Guinea coast of Africa, Fritillaria Meleagris, L.

HAG-BERRY see HEG-BERRY.

HAG-TAPER, G. *unholdenkerze*. Gerarde tells us that " Apuleius reporteth a tale of Ulysses, Mercurie, and the inchauntresse Circe using these herbes in their incantation and witchcrafts."   See HIGTAPER, in our Modern Floras incorrectly spelt *Hightaper*.   Verbascum Thapsus, L.

HAIR-BELL, an unauthorized but very plausible correction of the more usual spelling, *Harebell*, a name descriptive of the bell-shaped flowers and delicate stalks of the plant,   Campanula rotundifolia, L.

HAIR-GRASS, an imitation of its Latin name, Aira, L.

HALLELUJAH, the wood-sorrel, from its blossoming between Easter and Whitsuntide, the season at which the Psalms were sung which end with that word, those, namely, from the 113th to the 117th inclusive.   It bears the same name in German, French, Italian, and Spanish for the same reason.   There is a statement in some popular works, that it was upon the ternate leaf of this plant that St. Patrick proved to his rude audience the possibility of a Trinity in Unity, and that it was from this called *Hallelujah;* an assertion for which there is no ground whatever.

Oxalis Acetosella, L.

HALM or HAULM, A.S. *healm*, straw, Du. *helm* and *halm*, O.H.G. *halam*, Russ. *slama*, from L. *calamus*, Gr. καλαμος, Skr. *kalama*, its root *hal*, conceal, cover, from its early and general use as thatch.   Psamma arenaria.

HARD-BEAM, from the hardness of its wood, the hornbeam,   Carpinus Betulus, L.

HARD-GRASS,   Rottboellia incurvata, L.

HARD-HAY, G. *hartheu*, or as it is spelt in old writers, *harthau*, from its hard stalks,

Hypericum quadrangulare, L.

HARD-HEADS, from the resemblance of its knotty involucre to a weapon called a *loggerhead*, a ball of iron on a long handle,   Centaurea nigra, L.

HARDOCK, a word that occurs in the oldest editions of Shakspeare, in K. Lear (Act iv. sc. 4), but in later ones is wrongly replaced with *Harlock*.   It seems to mean the burdock, and to be so called from its involucres getting entangled in wool and flax, and forming the lumps called in old works *hardes* or *herdes*, which is explained by Batman on Bartholomew (c. 160), as "what is called in Latin *stupa*, and is the clensing (*i.e.* the refuse) of hempe and flexe," the equivalent of Fr. *bourre* from L. *burra*; as is evident from a passage in the Romaunt of the Rose, where Chaucer translates the phrase (l. 1233),

"Elle ne fut de *bourras*"

by

"That not of hempen *herdes* was."

Hardock will therefore be exactly equivalent to *Burdock*.

Arctium Lappa, L.

HARE-BELL, a name to which there is no corresponding one in other languages, in England assigned by most writers to Campanula rotundifolia, L. in Scotland, and in some English works, including Parkinson's Paradise, to the bluebell, Scilla nutans, Sm.

HARE'S-EAR, L. *auricula leporis*, from the shape of the leaves, Bupleurum rotundifolium, L. and also Erysimum orientale, L.

HARE'S-FOOT, Fr. *pied de lievre*, G. *hasenfuss*, from its soft downy heads of flowers, Trifolium arvense, L.

HARE'S LETTUCE, from its name in Apuleius, *Lactuca leporina*, called so, says he, because "when the hare is fainting with heat, she recruits her strength with it:" or as Anthony Askham says, "yf a hare eate of this herbe in somer, when he is mad, he shal be hole." Topsell also tells us in his Natural History, p. 209, that, "when Hares are overcome with heat, they eat of an herb called *Lactuca leporina*, that is the Hares-lettice, Hares-house, Hares-palace; and there is no disease in this beast, the cure whereof she does not seek for in this herb."

Sonchus oleraceus, L.

HARE'S PALACE, Fr. *palais de lièvre*, L. *palatium leporis*,

G. *hasen-haus*, the same as the hare's lettuce, and so called
from a superstition that the hare derives shelter and courage
from it; as we learn from the Ortus Sanitatis, ch. 334:
"Dises kraut heissend etlich hasenstrauch, etlich hasen-
hauss; dann so der hase darunder ist, so furchtet er sich
nit, und duncket sich gantz sicher, wann dises kraut hat
macht über die melancoley. Nun ist kein thiere als gar ein
melancholicus als der hase."

Sonchus oleraceus, L.

HARE'S-PARSLEY, in Aubrey's Wilts, probably

Anthriscus sylvestris, L.

HARE'S-TAIL, from its soft flower-heads,

Lagurus ovatus, L.

HARE'S-TAIL-RUSH, a translation of Lat. *Juncus cum
cauda leporina*, its name in Bauhin (Th. Bot. ii. 514),
and Plukenet (Alm. 201.), from the protrusion, after flowering, of soft hypogynous bristles resembling a hare's tail,
and its wiry rush-like stems,   Eriophorum vaginatum, L.

HARE-THISTLE, see HARE'S-LETTUCE.

HARIF, HEIRIFF, HAIREVE or HARITCH, in Pr. Pm.
hayryf, A.S. *hegerife*, from A.S. *hege*, hedge, and *reafa*,
which, significantly enough, means both a tax-gatherer and
a robber, so called, we may suppose, from its plucking wool
from passing sheep; originally the burdock, at present the
goose-grass,   Galium Aparine, L.

HARLOCK, as usually printed in K. Lear (a. iv. sc. 4), and
in Drayton, Ecl. 4:

"The honeysuckle, the *harlocke*,
The lily, and the lady-smocke:"

is a word that does not occur in the herbals, and which the
commentators have supposed to be a misprint for *charlock*.
There can be little doubt that *Hardock* is the correct reading
and that the plant meant is the one now called *Burdock*.
See above HARDOCK.   Arctium Lappa, L.

HARSTRONG, or HORESTRONG, Du. *harstrang*, G. *harn-strange*, strangury, from its supposed curative powers in this complaint, Peucedanum officinale, L.

HART'S CLOVER, the melilot, so called, says R. Turner (Bot. p. 199), "because deer delight to feed on it,"
Melilotus officinalis, L.

HART'S-HORN, from its furcated leaves,
Plantago Coronopus, L.

HART'S-THORN, Florio in v., the buckthorn, L. *spina cervina*, Rhamnus catharticus, L.

HART'S-TONGUE, from the shape of the frond, the *Lingua cervina* of the apothecaries,
Scolopendrium vulgare, Gärt.

HARTWORT, so called, because, as Parkinson tells us, (Th. Bot. p. 908), "Pliny saith that women use it before their delivery, to help them at that time, being taught by *hindes* that eate it to speade their delivery, as Aristotle did declare it before." Tordylium maximum, L.

HARVEST-BELLS, from its season of flowering,
Gentiana Pneumonanthe, L.

HASK-WORT, a plant used for the *hask* or inflamed trachea, being from its open throat-like appearance supposed, on the doctrine of signatures, to cure throat diseases. *Hask* in the Pr. Pm. is set down as synonymous with *harske*, austere, Sw. and Du. *harsk*, a term applied to fruits. Turner writes it *harrish*, as "dates are good for the harrishnes or rough-nes of the throte," or what we should at present call *huskiness.* Campanula latifolia, L.

HASSOCKS, A.S. *cassuc*, rushes, sedges and coarse grasses. "In Norfolk coarse grass which grows in rank tufts on boggy ground is termed hassock." A. Way in Pr. Pm. in v. "Hassock." The use of this term for the thick matted foot-stools used in churches seems to be taken from the application to such purpose of the natural tumps of a large sedge, the Carex paniculata, L.

HATHER, see HEATH.

HAVER, wild oat, Du. *haver*, G. *haber* or *hafer*, O.H.G. *haparo*, O.N. *hafra*, Sw. *hafre*, Da. *havre*, Wal. *hafar*, a name that, according to Holmboe, once meant corn generally, but was gradually restricted to the species most commonly used, the oat. J. Grimm (Gesch. d. D. Spr. i. 66,) supposes it to be related to L. *caper*, a he-goat, but Diez with more probability derives it from L. *avena*, with the usual prefix of an aspirate, and the change of *n* to *r*.

Avena sativa, etc., L.

HAWK-BIT, or HAWK-WEED, from a notion entertained by the ancients that with this plant hawks were in the habit of clearing their eyesight. See Pliny (l. xx. c. 7).

Hieracium, L.

HAWK'S BEARD, a name invented by S. F. Gray, and assigned, without any reason given, to the genus

Crepis, L.

HAWK-NUT, a name of which Ray says, (Syn. p. 209,) "cujus nominis rationem non assequor," but undoubtedly corrupted from *Hog-nut*, as it is correctly spelt in Jacob's Pl. Fav. p. 16. Bunium flexuosum, With.

HAWTHORN, the thorn of *haws*, *hays*, or *hedges*, A.S. *hagaðorn*, *hæg-*, or *hegeðorn*, G. *hagedorn*, Sw. *hagtorn*, an interesting word, as being a testimony to the use of hedges, and the appropriation of plots of land, from a very early period in the history of the Germanic races. The term *haw* is incorrectly applied to the fruit of this tree in the expression "hips and haws," meaning, as it does, the fence on which it grows, A.S. *haga* or *hæge*, G, *hage*.

Cratægus Oxyacantha, L.

HAYMAIDS, or HEDGEMAIDS, the ground ivy, a plant common in *hays* and hedges, which has derived the second syllable of its name from having been used as a "gill" to ferment beer, Fr. *guiller*, a word that also bore the meaning of "girl," or "maid," as in the proverb

"Every Jack must have his *Gill*." From the same equivocation have arisen other such names as "Lizzie, up the hedge!" etc. See GILL. Nepeta Glechoma, Benth.

HAZEL, A.S. *hæsl* or *hæsel*, and, allowing for dialect, the same word in all Germanic languages, the instrumental form of A.S. *hæs*, a behest, an order, from A.S. *hatan*, O.H.G. *haizan*, G. *heissen*, give orders, a hazel stick having been used to enforce orders among slaves and cattle, and been the baton of the master. J. Grimm (Gesch. d. Deuts. Spr. p. 1016,) observes, "Der hirt zeigt uns das einfache vorbild des fürsten, des ποιμην λαων, und sein *haselstab* erscheint wieder im zepter der könige : 'hafa i hendi *heslikylfo*' ['hold in hand a *hazel* staff'];" an expression that occurs in Sæmund's Edda in the second lay of the Helgaquida, str. 20. The verb *hælsian*, foretell, seems to be derived from the use of the hazel rod for purposes of divination. Corylus Avellana, L.

HAZEL-WORT, G. *hasel-wurz*, from the similarity of its calyx to the involucre of a nut, and not, as the books tell us, from its growing under hazel bushes. Frisch considers it to be corrupted from L. *asarum*. A. europæum, L.

HEADACHE, or HEAD-WARKE, from the effect of its odour, the red field-poppy, Papaver Rhœas, L.

HEART'S-EASE, a term meaning 'a cordial,' as in Sir W. Scott's Antiquary, ch. xi: "buy a dram to be eilding and claise, and a supper and *hearts-ease* into the bargain," given to certain plants supposed to be cardiac ; at present to the pansy only, but by Lyte, Bulleyn, and W. Turner to the Wallflower equally. The most probable explanation of the name is this. There was a medicine "good," as Cotgrave tells us, "for the passions of the heart," and called *gariofilé*, from the cloves in it, L. *caryophilla*. The wall-flower also took its name from the clove, and was called *giroflée*, from the same Latin word. See GILLIFLOWER. The cardiac qualities of the medicine were also extended to it, and the name of *Heart's-ease;* and, as the wallflower and the pansy were both comprehended among the Violets, that of *Hearts-ease* seems to have been transferred from the former to the species of the latter now called so. H. Brunschwygk, in his curious work "de arte distillandi," tells us of the wallflower: "Gel violen wasser kület ein wenig das herz: das geschycht uss ursach syner kreftigung und sterckung, ob es zu vil keltin het, so temperier es, ob es zu vil hytz het, so temperier es ouch darumb das *es das herz erfröwet.*" Tabernæmontanus also, (Kraüt. p. 689,) says of the wallflower: "Welchem menschen das Herz zittert von Kälte, der soll sich dieses gebrauchen." The instances are so numerous of the transference of an appropriate name to a plant to which it is quite unsuited, that we can find no difficulty in assigning this origin to the term *Hearts-ease* as at present employed.

Viola tricolor, L.

HEART-CLOVER, or, TREFOIL, "is so called," says W. Coles, in his Art of Simpling, p. 89, "not onely because the leaf is triangular like the heart of a man, but also because each leafe contains the perfect icon of an heart, and that in its proper colour, viz., a flesh colour. It defendeth the heart against the noisome vapour of the spleen."

Medicago maculata, Sibth.

HEATH, HEATHER, or HATHER, A.S. *hæð*, G. *heide*, O.N. *heiði*, Go. *haiþi*, a word which primarily meant the country in which the heath grows, Skr. *kshétra*, a field, Beng. *kheta*, and Skr. *kshiti*, land, from *kshi*, dwell. It is from the same root, *kshi*, that is derived Skr. *kshamá*, ground, Prakr. *khamá*, to which are related Gr. χαμαι, Go. *haims*, O.N. *heimi*, and our *home*. When the north of Europe was a forest, open land was naturally preferred for the site of dwellings, the heath was the only open land, and this acquired a name that had been used to designate a field, or homestead.

Erica and Calluna.

HEATH, IRISH-, or ST. DABEOC'S-,
> Menziesia polifolia, L.

„ SEA-, Frankenia pulverulenta, L.

HEATH-CYPRESS, from its resemblance to a small cypress tree, and its growth upon heathy ground,
> Lycopodium alpinum, L.

HEATH-PEA, from its pea-like esculent tuber, and usually growing upon sandy heaths,
> Lathyrus macrorrhizus, Wim.

HEDGE BELLS, G. *zaun-glocke*, a local, but expressive name for the larger bindweed, Convolvulus sepium, L.

HEDGEHOG PARSLEY, from its prickly burs,
> Caucalis daucoides, L.

HEDGE HYSSOP, a name transferred from a foreign species, a Gratiola, to the lesser skullcap,
> Scutellaria minor, L.

HEDGE-MAIDS, see HAYMAIDS.

HEDGE MUSTARD, Sisymbrium officinale, L.

HEDGE NETTLE, or -DEAD NETTLE, Stachys sylvatica, L.

HEDGE PARSLEY, Caucalis Anthriscus, H.

HEDGE TAPER, see HIG-TAPER.

HEDGE THORN, see HAWTHORN.

HEDGE VINE, the Virgin's bower, Clematis Vitalba, L.

HEG-BERRY, HEDGE-BERRY, HAG-, or HACK-BERRY, Sw. *hägg*, N. and Da. *hægebær*, "a wood berry," from a wood being in the northern counties called a *hag*, a word related to A.S. *hege*, hedge, Prunus Padus, L.

HELL-WEED, dodder, so called from the trouble and ruin it causes in flax fields. Threlkeld observes, that "after it has fastened upon a plant, it quits the root, and like a coshering parasite lives upon another's trencher, and first starves, and then kills its entertainer. For which reason irreligious clowns curse it by the name of *Hell-weed* and *Devil's-guts*." Cuscuta, L.

HELLEBORE, L. *helleborus*, Gr. ἐλλέβορος, a word of doubtful origin, at present applied to certain ranunculaceous plants, but not so in ancient works.

,,	BLACK-, the Christmas rose,	H. niger, L.
,,	FETID-, or STINKING-,	H. fœtidus, L.
,,	GREEN-,	H. viridis, L.
,,	WINTER-,	Eranthis hyemalis, S.

HELLEBORINE, from the resemblance of its leaves to those of a veratrum called "white hellebore," Epipactis, RB.

HELM, see HALM.   Psamma arenaria, PB.

HELMET-FLOWER, from the shape of the corolla,
Scutellaria, L.

HEMLOCK, or, as Gerarde spells it, HOMLOCK, A.S. *hœm* or *healm*, straw or haulm, and *leac*, plant, so called from the dry hollow stalks that remain after flowering ; a name originally applied to any of the Umbelliferæ, at present confined to two poisonous species.

,,	COMMON-,	Conium maculatum, L.
,,	WATER-,	Cicuta virosa, L.

HEMP, A.S. *hœnep*, It. *canapa*, Gr. κανναβις, from an Oriental root, Pers. *keneb*, Ar. *kinneb*, or *qunnab*, Arm. *ganap*. Herodotus speaks of the plant, (b. iv. c. 74) as a novelty lately introduced into Thrace from Scythia.
Cannabis sativa, L.

HEMP-AGRIMONY, from the resemblance of the leaves to those of *hemp*, and its being classed by the old Herbalists with *agrimony* under the general name of *Eupatorium*, or, as Gerarde writes it, *Hepatorium*,   E. cannabinum, L.

HEMP-NETTLE, or more properly HEMP-DEAD-NETTLE, from its flowers resembling those of the dead-nettles, and its leaves the leaflets of hemp,   Galeopsis Tetrahit, L.

HENBANE, a plant so called from the baneful effects of its seed upon poultry, of which Matthioli says (l. iv. c. 64) that "birds, especially gallinaceous birds, that have eaten the seeds perish soon after, as do fishes also." In old works it is called *Henbell,* A.S. *henne-belle,* a word that would seem to refer to the resemblance of its persistent and enlarged calyx to the scallop-edged bells of the middle ages, and the more so as the plant is called in some of the old plant-lists *Symphoniaca,* from *symphonia,* a ring of bells to be struck with the hammer. See below, YEVERING BELLS. Nevertheless this is possibly a case, such as so frequently occurs, of accommodating an ill-understood name to plain ideas. The plant is called in A.S. *belene,* and *belune,* in German *bilse,* in old German *belisa,* Pol. *bielún,* Hung. *belénd,* Russ. *belená,* words derived (according to Zeuss, p. 34) from an ancient Celtic God *Belenus* corresponding to the Apollo of the Latins : " Dem Belenus war das Bilsenkraut heilig, das von ihm *Belisa* und Apollinaris hiess." It is only in comparatively recent times that the *bell* has been replaced by *bane,* and it is difficult to see how it ever was connected with *henne,* if the *henne* referred to the bird so called.   Hyoscyamus niger, L.

HENBIT, G. *hüner-biss,* Fl. *hoender-beet,* L. *morsus gallinæ,* from some fancied nibbling of its leaves by poultry :

    „ THE GREATER-,   Lamium amplexicaule, L.

    „ THE LESSER-,   Veronica hederifolia, L.

HEN'S-FOOT, a mere translation of Lat. *pes pulli,* a name that Stapel, in Theophrast. p. 812, says was given to it from the resemblance of its leaves to a hen's claw, an observation that he must have made on a bad picture,

Caucalis daucoides, L.

HEPATICA, an adopted Greek word, adj. of *hepar,* the liver, applied to a plant with three-lobed leaves, used in affections of that organ : " *Hepatica* in Hepatis morbis, quod folia visceris istius gerunt figuram." Linn. Bibl. Bot. p. 117. In the medieval writers, such as Platearius, it means a Marchantia,   M. polymorpha, L. in popular language at the present day,

Anemone hepatica, L.

HERB BENNETT, L. *Herba benedicta*, Blessed Herb, the avens, so called, Platearius tells us, as quoted in Ort. San. c. clxxix, because " where the root is in the house the devil can do nothing, and flies from it: wherefore it is blessed above all other herbs." He adds that if a man carries this root about him, no venomous beast can harm him. The author of the Ortus says further, that, where it is growing in a garden, no venomous beast will approach within scent of it.   Geum urbanum, L.

„ also the hemlock from the same cause perhaps, since we learn from Ort. San. c. lxxxvii, that, on Pliny's authority, serpents fly from its leaves, because they also chill to the death, " sye auch kelten biss auf den tod." Pliny, however, alludes to a different plant from this.

Conium maculatum, L.

„ also the valerian, Sp. *yerva benedetta* as being a preservative against all poisons, Tab. i. p. 471 ; and therefore " gut fur die biss der bösen vergifftigen thieren," says Brunschwygk.   Valeriana officinalis, L.
But in point of fact the proper name of these plants was not *Herba benedicta*, but *S⁺ᵃ Benedicti herba*, St. Benedict's herb, G. *Sanct Benedicten-kraut*, and was assigned to such as were supposed to be antidotes, in allusion to a legend of St. Benedict, which represents, that upon his blessing a cup of poisoned wine, which a monk had given him to destroy him, the glass was shivered to pieces. (Mrs. Jameson, Mon. Ord. p. 9).

HERB CHRISTOPHER, a name vaguely applied to many plants which have no qualities in common. The three to which it has been given by our own writers are—
1st, the baneberry, "but," says Parkinson, "from what cause or respect it is called so, I cannot learn,"

Actæa spicata, L.
2nd, the Osmund fern,   Osmunda regalis, L.
3rd, the fleabane,   Pulicaria dysenterica, Cass.

But besides these there are others which by the older herbalists are called *S^{tt.} Christophori herba*, and *Kristoffels-kraut;* such as Spiræa ulmaria, L. Gnaphalium germanicum, Hud. Betonica officinalis, L. Vicia Cracca and sepium L.

HERB GERARD, Du. *Geraerts cruyt*, called so, says Forster in Nat. Phen. p. 101, from St. Gerard, who used to be invoked against the gout,

Ægopodium Podagraria, L.

HERB OF GRACE, rue, from this word *rue* having also the meaning of "repentance," which is needful to obtain God's grace, a frequent subject of puns in the old dramatists. See quotations in Loudon's Arboretum, vol. i. p. 485.

Ruta graveolens, L.

HERB IMPIOUS, from the younger flowers overtopping the older ones, like undutiful children rising over the heads of their parents; "ob id impiam appellavere, quoniam liberi super parentem excellant." Plin. Nat. Hist. (l.xxiv. 113).

Gnaphalium germanicum, Huds.

HERB IVY, or HERB IVE, or -EVE, a name given to several different plants with deeply divided leaves, a corruption of the *Abiga* of Pliny. Beckmann, in his Lexicon Botanicum, explains it thus: "*Iva*, Ruellius *ibiga;* hinc duabus abjectis literis *i* et *g*, *iba*, et tandem *iva* manavit in vulgi nomenclationem." The name has been given by the herbalists to　　　　Ajuga Iva, Coronopus Ruellii,

and Plantago Coronopus, L.

HERB OF LIFE, in Erasmus' Praise of Folly, some mythical plant that cannot be identified.

HERB MARGARET, the daisy, see MARGUERITE.

Bellis perennis, L.

HERB PARIS, incorrectly so spelt with a capital P, being its Lat. name *Herba paris*, Herb of a pair, of a betrothed couple, in reference to its four leaves being set upon the stalk like a trulove-knot, the emblem of an engagement, whence its synonym, *Herb Trulove,*

Paris quadrifolia, L.

HERB PETER, the cowslip, from its resemblance to St. Peter's badge, a bunch of keys, whence its G. name, *schlüssel-blume,*   Primula veris, L.

HERB ROBERT, a geranium, that according to Adelung, was so called from its being used to cure a disease known in Germany as the *Ruprechts-Plage,* from Robert, duke of Normandy, for whom was written the celebrated medical treatise of the middle ages, the "Regimen Sanitatis Salernitanum," or "Schola Salernitana." In some old writers it is called *Sancti Ruperti herba.* See Bauhin de plant. sanct. p. 81. There are four saints of that name, and one of them, a bishop of Salzburg, is said by Alban Butler in his Lives of the Saints to have cured all the maladies of body and soul. In a MS. vocabulary of the 13th century, in Rel. Ant. vol. i. p. 37, it occurs as "*Herba Roberti,* Herb Robert." Adelung's explanation is probably the correct one.   Geranium Robertianum, L.

HERB TRINITY, L. *Herba Trinitatis,* G. *dreifaltigkeits-blume,* from having three colours combined in one flower, the pansy,   Viola tricolor, L. and also from having three leaflets combined into one leaf,

Anemone hepatica, L.

HERB TWOPENCE, from its pairs of round leaves, the moneywort,   Lysimachia Nummularia, L.

HERB WILLIAM, in R. Turner's Botanologia, p. 45,

Ammi majus, L.

HERON'S BILL, from the shape of the seed vessel,

Erodium, L'Her.

HIG-TAPER, HAG-TAPER, in Turner, HYGGIS-TAPER, in Lupton, i. 3, HEDGE-TAPER, incorrectly spelt in our floras *High-taper,* either from A.S. *hig* or *hyg,* hay and *taper,* meaning a taper made of dry herbage, or from A.S. *hege* or *haga*, a hedge, the usual place of its growth. Dodoens tells us that it was called *candela*, " folia siquidem habet mollia, hirsuta, ad lucernarum funiculos apta ;" "a plant," says the Grete Herball, "whereof is made a maner of lynke, if it be talowed ;" or, as Parkinson says, (Th. Bot. p. 62,) "Verbascum is called of the Latines *Candela regia* and *Candelaria*, because the elder age used the stalks dipped in suet to burne, whether at Funeralls or otherwise." Brunfelsius, ed. 1531, says, p. 197, that it is called *Wull-* or *König-kerz*, "darum das sein stengel von vilen gedörrt würt, überzogen mit harz, wachs, oder bech, und stangkerzen oder dartschen davon gemacht, und gebrannt für schaub-fackelen." *Schaub* means a wisp of straw, and rather supports the above explanation of *hig* as hay.

Verbascum Thapsus, L.

HINDBERRY, A.S. *hindberie*, a name that was once very generally given to the raspberry, and is still retained in some counties, derived, apparently, from *hind*, as the gentler, the tamer kind of bramble, contrasted with the *heorot-berie*, the hart-berry. Adelung suggests its derivation from Lat. *Idæus*, by a change that is quite consistent with analogy, viz. prefixing an *h* to the initial vowel, and an *n* before *d*. Ray's explanation, " forte sic dicta, quia inter hinnulos et cervos, i.e. in sylvis et saltibus. crescunt," is very unsatisfactory, a mere guess, as are nearly all his explanations of English names.

Rubus idæus, L.

HINDHEAL, A.S. *hind-hele*, *-heoleðe*, or *-hæleð*, from its curing the hind, probably the same herb as the *Elaphoboscum*, which deer were supposed to eat when stung by serpents, and of which Lupton tells us, No. 80, that "it is said that harts in Crete being struck with darts envenomed, do eat of a certain herb called Dittany, and thereby the prick that sticks in them, is driven out." Teucrium Scorodonia, L.

HIP-, or HEP-ROSE, the dog-rose, that which bears the *hip*, A.S. *hiop, heap, heope*, O.S. *hiopa*, O.H.G. *hiofa*, Norw.

hiupa, and *jupe*, a corruption from Gr. ζιζυφον, M. Lat. *jujuba*, the *j* of which has become *hi*, as in many other instances, through the intermediate sound of *y*. Thus *Job* becomes in German *Hiob*, and *jejunium*, *hunger*, and conversely an initial *hi* is in Norway always pronounced *y ;* *hierte*, heart, *yerte;* and similarly in our own western counties *heifer* becomes *yeffer*. It seems to confirm this view of the derivation of *hip* from *jujuba*, that the Ortus Sanitatis, the figures of which, bad as they are, are traditional copies of very ancient ones, gives (c. ccxx), a rosebush with hips on it for the *Jujube*, and titles the chapter: "Hanbotten, jujube grece et latine:" the *hanbotte* being the same as the *hagebutte* or hip. In the Old Saxon of the Heliand, *hiopa* seems to have meant the briar rather than the fruit: (l. 3488) "nec oc figun ni lesat an *hiopon :*" nor gather figs on hips : where *hiopon* represents the Gr. τριβολος of Matth. vii. 16, a word that Wycliffe translates "breris," briars, but our authorized version, less correctly, " thistles."

Rosa canina, L.

HIP-WORT, from the resemblance of the leaf to the acetabulum or hip-socket, whence its former name of *herba coxendicum*, herb of the hips,   Cotyledon umbilicus, Hud.

HIRSE, G. *hirse*, L. *cererisia*, ale, from ale being brewed from it, a kind of millet,   Panicum, L.

HOCK, or HOCK-HERB, the mallow, from Lat. *Alcea* by the change of *l* to *u*, and the usual prefix of *h* to Lat. words beginning with a vowel upon their becoming English ; *Alc, auc, hauc.* See HOLLIHOCK.   Althæa and Malva, L.

HOG'S-FENNEL, a coarse rank plant, fennel for a hog,

Peucedanum officinale, L.

HOG-NUT, the pig-nut,   Bunium flexuosum, With.

HOG-WEED, from the fondness of hogs for its roots, the cow-parsnep,   Heracleum Sphondylium, L.

HOLLIHOCK, in Huloet's Dict^y. HOLY HOKE, a perplexing word. The *hock* is clearly from the L. *alcea*, (see

HOCK,) but the *Holli* is very difficult to explain. The most probable origin of it is L. *caulis*, with the meaning of a *cale-, coley-*, or *cabbage-hock*, and referring, as in *cabbage-rose*, to its well-filled double flówers, or used in the sense of *stalk*, and referring to its lofty habit, in contrast with that of the lowly *Hock-herb*, or mallow. *Cauli-* or *Coley-hock* would easily pass into *Holly-* and *Holy-hock*.

Althæa rosea, L. and ficifolia, Cav.

HOLLOW-WORT, or HOLE-WORT, from its hollow root,

Corydalis tuberosa, DC.

HOLLY, or HOLM, on the Eastern Border called HOLLEN, the old form of the word, and that from which *holm* has been formed by the change of *n* to *m*, as Lime from Line ; A.S. *holen* or *holegn*, a word derived from L. *ulex*, which in the middle ages was confused with *ilex*, the holm oak of the ancients, whence the adjective *uligna*, and with the aspirate, *huligna* and *holegn*. The form *Holly* will have been the more readily adopted, and the tree have been called in Anglo-Saxon times *elebeam*, or oil-tree, from its branches having being used for *olive* branches, and strewed before the image of Jesus, in certain solemnities of the church that represented his entrance into Jerusalem. Thus in Googe's Naogeorgus :

> " He is even the same, that, long agone,
>   While in the streete he roade,
>   The people mette, and *olive* bowes
>   So thicke before him stroade."

Ilex Aquifolium, L.

„ KNEE-, Ruscus aculeatus, L.

„ SEA-, Eryngium maritimum, L.

HOLLY-, or HOLM-OAK, see HOLLY. Quercus Ilex, L.

HOLY GHOST, so called "for the angel-like properties therein," says Parkinson, (Th. Bot. p. 941). " It is good against poison, pestilent agues, and the pestilence itself," says W. Turner (b. iii. 5.). Angelica sylvestris, L.

HOLY GRASS, from its Gr. name, *ιερα χλοη,*

Hierochloe borealis, Rm.

HOLY HAY, in some old works, the lucern, a mistaken translation of Fr. *sain-foin.* See SAINFOIN.

Medicago sativa, L.

HOLY HERB, L. *herba sacra,* Gr. *ιερα βοτανη,* the vervain, so called, says Dioscorides, (b. iv. 61,) *δια το εύχρηστον έν τοις καθαρμοις έιναι εις περιαμματα,* "from its being good in expiations for making amulets." It acquired this character from being used to decorate altars : " Ex ara sume *verbenas* tibi ;" Ter. in Andria ; and, as Pliny tells us (b. 22, c. 2) *" "non aliunde sagmina in remediis publicis fuere, et in sacris legationibus, quam *verbena ;*" and adds, "hoc est gramen ex ara cum sua terra evulsum." It would seem that branches of any kind used about the altar at sacred festivals were called *verbena,* and being borne by an ambassador rendered his person inviolable ; and that the word did not originally apply exclusively to that which we now call *vervain.* Verbena officinalis, L.

HOLY ROPE, a plant that from its hemp-like leaves was fixed upon as the one that yielded the rope with which Jesus was bound; just as there was a Christ's thorn, a Christ's gall, a reed-mace, a Christ's ladder, etc., found to represent the other incidents of the Crucifixion,

Eupatorium cannabinum, L.

HONESTY, from the transparency of its dissepiments,

Lunaria biennis, L.

HONEWORT, from its curing the *hone,* a hard swelling in the cheek so called, Ger. p. 1018, Trinia glaberrima, L.

and Sison Amomum, L.

HONEYSUCKLE, A.S. *hunig-sucle,* a name that is now applied to the woodbine, but it is very doubtful to what plant it properly belongs. In the A.S. vocabularies it is translated *Ligustrum,* which in other places means the cowslip and primrose. In Parkinson and other herbalists it is assigned to the meadow clover, which in our Western counties is still called so. This, too, is the plant meant by the *rede hony suckle gres* in the Stockholm Med. M.S. (Archæol. v. xxx. p. 399). The name seems to have been transferred to the woodbine on account of the honey-dew so plentifully deposited on its leaves. But it is not at all clear, what is the proper meaning of the word *honeysuckle*. In poetry and popular usage, Lonicera, L. but in farmer's language, the meadow clover,

Trifolium pratense, L.

HONEYSUCKLE, DWARF-, Cornus suecica, L.

   ,, FLY-, Lonicera xylosteum, L.

   ,, FRENCH-, a plant used on the Continent for forage as the meadow clover is with us; a foreign honeysuckle-clover ; Hedysarum coronarium, L.

HONEY-WARE, A.S. *war*, sea-weed, and *honey* from its being covered with a layer of sugar, " dont les Islandais se servent très bien," says Duchesne, p. 364. Alaria esculenta, and Laminaria saccharina, Lam.

HOODED MILFOIL, Fr. *millefeuille*, thousand leaf, from its very finely divided leaves and the hood shape of its corolla, Utricularia vulgaris, L.

HOOK-HEAL, from its being supposed on the doctrine of signatures to heal wounds from a bill-hook, which its corolla was thought to resemble, Prunella vulgaris, L.

HOP, a name adopted from the Netherlands with the culture of the plant, L. Ger. *hoppen*, G. *hopfe*, M. Lat. *hupa*, possibly connected with words that mean " head," as *haupt*, *haube*, etc. but only by an accidental coincidence approaching the Fr. *houblon*, Humulus Lupulus, L.

HOP-CLOVER, from the resemblance of its fruiting capitules to little heads of hop,

Trifolium procumbens, and agrarium, L.

HOREHOUND, A S. *hara-hune*, from *hara*, hoary, and *hune*, honey, a name that may have been, and most likely was, suggested by that of another labiate plant, *melissa*, Gr. μελισσα, honey, but may very possibly be a corruption of Lat. *Urinaria;* the plant having been regarded as one of great efficacy in cases of strangury and dysuria.   See Ort. Sanit. ch. 256.   Marrubium vulgare, L.

HOREHOUND, BLACK-, from its dark flowers,
                              Ballota nigra. L.

  „   WATER-,   Lycopus europæus, L.

HORE-STRANGE, or -STRANG, from its supposed virtue in strangury.   See HARSTRONG.

HORNBEAM, or -BEECH, from its wood being used to yoke horned cattle, " as well by the Romans in old time," says Gerarde, p. 1479, " as in our own, and growing so hard and tough with age as to be more like horn than wood,"
                              Carpinus Betulus, L.

HORNWORT, from its bi- and tri-furcate leaves,
                              Ceratophyllum, L.

HORNED POPPY, from its long curved horn-like seedpods,
                              Glaucium luteum, L.

HORSE-BANE, from its being supposed in Sweden to cause in horses a kind of palsy, an effect that has been ascribed by Linnæus not so much to the noxious qualities of the plant itself, as to an insect, *curculio paraplecticus*, that breeds in its stem (Syst. Nat. 610),
                              Œnanthe Phellandrium, Lam.

HORSE-BEAN, the variety of bean grown for the food of horses,   Vicia Faba, L.

HORSE-BEECH, see HURST-BEECH, of which it is a corruption.

HORSE-CHESTNUT, said to be called so from its fruit being used in Turkey, the country from which we received it, as food for "horses that are broken or touched in the wind;" see Selby, p. 34.   Parkinson says (Th. Bot. p. 1402): " Horse chesnuts are given in the East, and so through all Turkie, unto Horses to cure them of the cough, shortnesse of winde, and such other diseases."   In this country horses will not eat them, and the name is more likely to have been given to these nuts to express coarseness. The ingenious conjecture of a writer in N. and Q. (3, ser. x. 45,) that it was suggested by the cicatrix of its leaf resembling a horse-shoe, with all its nails evenly placed, has no support of ancient authors. Æsculus Hippocastanum, L.

HORSE-CHIRE, the germander, from its growing after horse-droppings, Fr. *chier,* Teucrium Chamædrys, L.

HORSE-FLOWER, Lyte, b. ii. c. 14, from Flem. *peerd-bloeme,* horse-flower, a name that it seems to have acquired from the *pyrum* of its Latin name having been misunderstood to mean " pear," Flem. *peere,* and this word being confused with *peerd,* a horse, Melampyrum sylvaticum, L.

HORS-HELE, -HEAL, or -HEEL, A.S. *hors-elene,* L. *Inula Helenium,* which by a double blunder of *Inula* for *hinnula,* a colt, and *Helenium* for something to do with *heels* or *healing,* has been corrupted into *Hors-hele,* and the plant employed by apothecaries to heal horses of scabs, and sore heels, Inula Helenium, L.

HORSE-HOOF, from the shape of the leaf,
Tussilago Farfara, L.

HORSE-KNOB, a coarse *knap*weed, Centaurea nigra, L.

HORSE-MINT, Mentha sylvestris, L.

HORSE-MUSHROOM, from its size as compared with the species more commonly eaten, Agaricus arvensis, Sch.

HORSE-PARSLEY, from its coarseness as compared with smallage or celery, Smyrnium Olus atrum, L.

HORSE-RADISH, Cochlearia Armoracia, L.

HORSE-SHOE-VETCH, Off. L. *ferrum equinum,* from the shape of the legumes, Hippocrepis comosa, L.

HORSETAIL, L. *cauda equina,* a name descriptive of its shape, and under which it was sold in the shops,
Equisetum, L.

HORSE-THYME, a coarse kind of thyme,
Calamintha Clinopodium, Benth.

HOUND'S-BERRY-TREE, or HOUND'S-TREE, a mistaken equivalent for *Dogwood*; see DOGWOOD.

HOUND'S-TONGUE, from the Gr. κυνογλωσσον now applied to a plant, of which W. Coles tells us in his Art of Simpling, ch. xxvii, that "it will tye the Tongues of Houndes, so that they shall not bark at you, if it be laid under the bottom of your feet, as Miraldus writeth." The name was probably given to the Greek plant on account of the shape and soft surface of the leaf, and in contrast to the rough bugloss or oxtongue.

Cynoglossum officinale, L.

HOUSE-LEEK, a leek or plant, A.S. *leac*, that grows on houses,   Sempervivum tectorum, L.

HOVE, A.S. *hufe*, a chaplet, after its Latin name, *corona*, Gr. στεφανωμα, and so called, says Parkinson, "because it spreadeth as a garland upon the ground," the ground ivy, also called in old MSS. Heyhowe, Heyoue, Haihoue, Halehoue, and Horshoue. See Pr. Pm. p. 250, note by Way.   Nepeta Glechoma, Benth.

HULST, Du. *hulst*, which Weiland derives from L. *ilex*, but without accounting for the terminal *st*, which would seem in this, and several other words, to be the Lat. *cetum*, indicating the locality of its growth. Its immediate origin would thus be *ulicetum* for *ilicetum*, a bed of *ulex* or *ilex*, two names frequently confused in medieval writings, whence also G. *huls*, and Fr. *houx*.   Ilex Aquifolium, L.

HULVER, in Chaucer HULFEERE, Fr. *olivier*, olive-tree, a name given to the holly from its being strown on the road in place of *olive* branches at the public festivals of the church; as was that of "palm," for a similar reason, to the flowering branches of the willow. See quotation under HOLLY. ·   Ilex Aquifolium, L.

  ,,   KNEE-, the butcher's broom, see under KNEE-.
  ,,   SEA-, from its prickly leaves,
                              Eryngium maritimum, L.

HUNGER-GRASS, from its starving cereal crops among which it grows, and thus causing famine,

> Alopecurus agrestis, L.

,, -WEED, from its abundance indicating a bad crop, and season of famine, Ranunculus arvensis, L.

HURR-BURR, the burdock, whose involucres form the nucleus of the *hardes* or *hurds*, Norw. *hörr*, the tangled lumps that are carded out of flax and wool,

> Arctium Lappa, L.

HURST- or HORST- or HORSE-BEECH, the hornbeam, from its growth on hursts, and some resemblance of its leaves to those of the beech-tree, Carpinus Betulus, L.

HURT-SICKLE, "because," says Culpeper, "with its hard wiry stem it turneth the edge of the sickle, that reapeth the corn;" called, for the same reason, by Brunsfelsius *Blaptisecula*, from Gr. βλαπτω, injure,

> Centaurea Cyanus, L.

HURTLE-BERRY, and HUCKLE-BERRY, corruptions of *Whortle-berry*, itself a corruption of *Myrtle-berry*,

> Vaccinium Myrtillus, L.

HYACINTH, Gr. ὑακινθος, a plant to which frequent allusion is made by the Greek poets, but which, from the vague way in which they used the names of flowers, it is impossible to identify. It can scarcely have been the hyacinth of our gardens. Some suppose it to have been the martagon lily, some a gladiole. The former seems to have been Ovid's plant, the latter that of the Sicilian poets, Theocritus and Moschus. But it would here be out of place to enter into the question. See Ovid. Met. b. x, 164. Theocr. Id. x, 28. Moschus, Id. iii, 6. As now understood, it is the genus Hyacinthus, L.

HYSSOP, the name given in our authorized version of the Bible to the caper, but in popular language assigned to a labiate plant, the supposed ὑσσωπος of Dioscorides,

> Hyssopus officinalis, L.

ICELAND MOSS, a lichen so called from its abundance in Iceland, whence it is imported for medicinal and culinary purposes, Cetraria Islandica, Ach.

INUL, L. *Inula*, see ELECAMPANE.

IREOS, the genitive case of *iris*, used by apothecaries to mean the orrice root, *radix* being understood. See ORRICE.

IRISH HEATH, from its occurrence chiefly in the west of Ireland, Menziesia polifolia, Jus.

IRISH, or CARRAGEEN MOSS, a sea-weed so called, imported from Ireland, Chondrus crispus, Lyngb.

IRON-HEADS, from the resemblance of its knobbed involucre to a weapon with an iron ball fixed to a long handle, called a Loggerhead, Centaurea nigra, L.

IRON-WORT, a translation of its Lat. name, *Sideritis*, from Gr. σιδηρον, iron, a name formerly applied to several different plants, supposed to heal wounds from iron weapons, but now confined to a genus of Labiatæ, of which we have no British representative. In Jacob's Plant. Faversh. Galeopsis Ladanum, L.

IVRAY, Fr. *ivraie*, drunkenness. See RAY-GRASS.

IVY, in MS. Sloane, No. 3489, 3, spelt *Icyne*, A.S. *ifig*, a word strangely mixed up with the names of the yew-tree, O.H.G. *ëbah*, from which, according to Grimm, arose *ëbowe, ëbhowe, ebihowe, ephou, epheu*, and in Alsace *epphau*. It seems to have originated with the Lat. *abiga*, used by Pliny as the name of the plant called in Greek Chamæpitys, and miswritten by some copyist *ajuga*, which was further corrupted to the M. Lat. *iva*. See YEW. Looking at the names of the two trees, the Ivy and the Yew, in the different languages of Europe, we cannot doubt that they are in reality the same word. Indeed in Höfer's "Wörterbuch der in Oberdeutschland üblichen Mundart" we find that *Ive* or *Ivenbaum* belongs equally to one or the other. In English we get *ivy* from *iva*, and *yew* from the same word, written *iua*. The source of the confusion seems to have been this. The Chamæpitys of Pliny, as we learn from Parkinson (Th. Bot. p. 284,) was "called in English *Ground-pine,* and *Ground-ivie,* after the Latin word *Iva.*" But this name *Ground-ivy* had been assigned to another plant, which was called in Latin *Hedera terrestris,* and thus *Ivy* and *Hedera* came to be regarded as equivalent terms. But there was again another plant that was also called *Hedera terrestris,* viz., the creeping form of Hedera helix, and as *Ivy* had become the equivalent of *Hedera* in the former case, so it did in this too, and eventually was appropriated to the full-grown ever-green shrub so well-known. How *iva* became the name of the Yew-tree will be explained below.   Hedera Helix, L.

IVY, GROUND-, see GROUND-IVY.

IVY-LEAFED CHICKWEED,   Veronica hederifolia, L.

JACINTH, Fr. *jacinthe,* L. *hyacinthus,* the hyacinth.

JACK BY THE HEDGE, from *Jack* or *Jakes,* latrina, alluding to its offensive smell, and its usual place of growth,
                      Alliaria officinalis, L.

JACK OF THE BUTTERY, a ridiculous name that seems to be a corruption of *Bot-theriacque* to *Buttery Jack,* the plant having been used as a theriac or anthelmintic, and called *Vermicularis* from its supposed virtue in destroying bots and other intestinal worms,   Sedum acre, L.

JACOB'S LADDER, usually supposed to be called so from its successive pairs of leaflets,   Polemonium cæruleum, L.

JASMINE, JESSAMINE, JESSE, or GESSE, Sp. and Fr. *jasmin,* from Pers. *jásemín,* Ar. *jásamûm,*
                      Jasminum, L.

JERSEY LIVELONG, from its occurrence in Jersey, and being of the same genus as many of the so-called everlasting flowers, ,   Gnaphalium luteo-album, L.

JERUSALEM ARTICHOKE, called *artichoke* from the flavour of its tubers, and *Jerusalem* from It. *girasole,* turn-sun, that is, a sun that turns about, the sun-flower, to which genus this plant belongs. By a quibble on *Jerusalem* the soup made from it is called " Palestine."

Helianthus tuberosus, L.

JERUSALEM COWSLIP, from being like a cowslip " floribus primulæ veris purpureis," as described by Lobel, and from having been confounded under the name of *Phlomis* with the Sage of Jerusalem, Pulmonaria officinalis, L.

JERUSALEM CROSS, from an occasional variety of it with four instead of five petals, of the colour and form of a Jerusalem cross, Lychnis chalcedonica, L.

JERUSALEM, OAK OF-, called *oak* from the resemblance of its leaf in an outline picture to that of the oak, and its confusion under the name of χαμαιδρυς, ground oak, with the Chenopodium Botrys. The *Jerusalem* seems here as in other cases to stand as a vague name for a distant foreign country. Teucrium Botrys, L.

  „ STAR OF-, It. *girasole*, turn-sun, in allusion to the popular belief that it turns with the sun, whence it was also called *solsecle*, from Lat. *solsequium*, A.S. *sol-sece*, (see JERUSALEM ARTICHOKE;) and *star* from the stellate expansion of the involucre. See SALSIFY.

Tragopogon porrifolius, L.

JEWS-EARS, L. *auricula Judæ*, a fungus resembling the human ear, and usually growing from trunks of the elder, the tree upon which the legend represents Judas as having hanged himself. Mandeville tells us that he saw the very tree. Exidia auricula Judæ, Fries.

JOAN-, or JONE-SILVER-PIN, the red poppy, called so, as Parkinson tells us, p. 367, because it is " Fair without and foul within," alluding to its showy flower and staining yellow juice. According to Forby the term " Joan's silver pin " means among the East Anglians " a single article of finery produced occasionally and ostentatiously among dirt and sluttery," and in this sense too it is a fit name for this

gaudy flower so conspicuous among the weeds of the corn-
field.   Papaver Rhœas, L.

JOINTED CHARLOCK, from its pod being contracted between the seeds, so as to appear articulated, but being
otherwise like charlock,   Raphanus Raphanistrum, L.

JONQUIL, Sp. *junquillo*, dim. of *junco*, L. *juncus*, a rush,
from its slender rush-like stem,   Narcissus Jonquilla, L.

JOSEPH'S FLOWER, Du. *Joseph's bloem*, in allusion to its
popular name of *Go-to-bed-at-noon*, and Joseph's refusal to
do so, the implements and memorials of their achievements
and trials having been usually adopted as the badge of
the Bible heroes, and exhibited in their pictures. Thus
Samson had his jawbone, David his giant's head, Jonah his
gourd, etc.   Tragopogon pratensis, L.

JOUBARB, see JUPITER'S BEARD.

JUDAS TREE, from this, and not the elder, being the one
upon which some of the legends represent that traitor as
having hung himself,   Cercis Siliquastrum, L.

JULY FLOWER, used by Drayton under the mistaken
notion that it gave the meaning of *Gilliflower*,

                                  Matthiola, L.

JUNIPER, from the Lat. *Juniperus*, a word of uncertain
origin,   J. communis, L.

JUNO'S ROSE, L. *rosa junonis*, the white lily. We are
told by Cassianus Bassus in his Geoponica (l. xi. c. 20),
that Jupiter, to make his son Hercules immortal, put him
to the breast of Juno while she was sleeping, and that the
milk which was spilt, as the child withdrew from her,
formed the milky way in the heavens, and gave rise to the
lily upon earth. The story is curious as an instance of one
that was transferred from the pagan to the Roman Catholic
mythology; for it is evidently the source of a similar one
of the Virgin and the milk thistle. See Lyte, b. ii. c. 42.
Lobel, Kruydtb. p. 201. Dodoens, ii. 2. 1.

                                  Lilium candidum, L.

JUNO'S TEARS, Gr. ʻHρας δακρυον, a name that by Dioscorides was given to the Coix lacryma, now called "Job's tears," in allusion to its hard, polished, bead-like seeds, but through some confusion with the *peristereon*, as in Apuleius, c. 66, transferred to the vervain, which has nothing about it that resembles a tear, Verbena officinalis, L.

JUPITER'S BEARD, Fr. *joubarb*, L. *Jovis barba*, the house-leek, so called from its massive inflorescence, like the sculptured beard of Jupiter.

> Quam sempervivam dicunt, quoniam viret omni
> Tempore : *barba jovis* vulgari more vocatur:
> Esse refert similem prædictæ Plinius istam.   Macer.
>
> Sempervivum tectorum, L.

JUPITER'S STAFF, R. Turner, Bot. p. 216, the mullein,
Verbascum Thapsus, L.

JUR-NUT, Da. *jord-nöd*, earth-nut,
Bunium flexuosum, L.

KALE, see CALE.

KATHARINE'S FLOWER, from the persistent styles spreading like the spokes of a wheel, the symbol of St. Katharine, from her having been martyred upon a wheel,
Nigella damascena, L.

KECKS, KEX, KECKSIES, KAXES, KIXES, or CASHES, the dry hollow stalks or haulms of umbelliferous plants, so called from an old English word *keek* or *kike* retained in the Northern counties (Brockett) in the sense of "peep" or "spy," Go. *kika*, Da. *kige*, Du, *küken*, a name suggested by their most obvious peculiarity : viz., that one may look through them. It is most commonly given to the stems of the cow-parsnep and cow-parsley.

KEDLOCK, or KETLOCK.   See CHEDLOCK.

KELPWARE, a sea-wrack or ware that produces kelp or barilla, Fucus nodosus, and vesiculosus, L.

KEMPS, the flower stalks of the ribwort plantain, A.S. *cempa*, a warrior, Da. *kæmpe*, a word peculiar to the Northern counties, and more probably of Danish than Ang. Sax. origin, alluding to the child's pastime of fighting them against one another, a game that is known in Sweden also, where they are called *Kämpar*, Plantago lanceolata, L.

KERNEL-WORT, from having kernels or tubers attached to the roots, and being therefore supposed, on the doctrine of signatures, to cure diseased *kernels*, or scrophulous glands in the neck, Scrophularia nodosa, L.

KIDNEY-VETCH, because " it shall prevayle much against the strangury, and against the payne of the reynes." Lyte, (b. i. ch. 7). Anthyllis vulneraria, L.

KIDNEY-WORT, from a distant resemblance of its leaves to the outline of a kidney, Umbilicus pendulinus, DC.

KING'S CLOVER, from its M.Lat. name *corona regia*, royal crown, "because," as Parkinson says, (Th. Bot. p. 720,) " the yellowe flowers doe crown the top of the stalkes," as with a chaplet of gold, Melilotus officinalis, L.

KING'S CUP or COB, AS. *copp*, a head, from the resemblance of the unexpanded flower-bud, and of its double variety, to a stud of gold, such as *kings* wore, Fr. *bouton d'or*, Ranunculus acris and bulbosus, L.

KING'S KNOB, see KING'S CUP. A.S. *cnæp*, a button, Da. *knap*.

KIPPER or KNIPPER NUT, called in Scotland *knapparts*, from *knap*, a knob, and *urt*, wort, the heath-pea, from its knotty tubers, Vicia Orobus, DC.

KISS-ME-ERE-I-RISE, KISS-ME-BEHIND-THE-GARDEN-GATE, LOOK-UP-, or JUMP-UP-AND-KISS-ME, see PANSY.

KISS-ME-TWICE-BEFORE-I-RISE, R. Turner. Bot. p. 223, the fennel flower, Nigella damascena, L.

KNAP-BOTTLE, from its inflated calyx, resembling a little bottle, and snapping when suddenly compressed, Du. *knappen*, crack, snap, Silene inflata, L.

KNAP-WEED, KNOP-, or KNOB-WEED, Da. *knopurt*, from its knob-like heads, A.S. *cnæp*, L.Ger. *knoop*, Da. *knop*, G. *knopf*,   Centaurea nigra, L.

KNAWEL, G. *knauel* or *knäuel*, a hank of thread, from its spreading stems,   Scleranthus, L.

KNEE-HOLM, -HULVER, or -HOLLY, A.S, *cneow-holen*, a shrub referred to the holms or hollies on account of its evergreen prickly leaves, and deriving the prefix *knee* from confusion, under the name of *daphne*, and *victoriola*, with Lat. *cneorum*, a plant used in chaplets, as were some species of this genus. See Pliny, N.H. xxi. 9.
                    Ruscus aculeatus, L.

KNIGHT'S SPURS, the larkspur, from its long, slender, projecting nectaries,   Delphinium, L.

KNIGHT'S-WORT, -WOUND-WORT, or -PONDWORT, from its sword-like leaves,   Stratiotes aloides, L.

KNIT-BACK, L. *confirma*, from being used as a strengthener or restorative, the comfrey,
                    Symphytum officinale, L.

KNOB-TANG, Da. *tang*, sea-weed, and *knob*, A.S. *cnæp*, a word connected with many others beginning with *kn*, *kl*, *gn*, and *gl*, in all the Germanic languages, and signifying a lump, or something knotted and hard,
                    Fucus nodosus, L.

KNOLLES, turnips, Du. *knol*, Da. *knold*, a tuber,
                    Brassica Rapa, L.

KNOT-BERRY, from the knotty joints of the stems,
                    Rubus Chamæmorus, L.

KNOT-GRASS, or KNOT-WORT, the centinode, from its trailing jointed stems and grass-like leaves. The "hindering knotgrass" of Shakspere, (M.N.D. iii. 2), was probably so called from the belief that it would stop the growth of children, as in Beaumont and Fletcher's Coxcomb, A. ii.:
"We want a boy
Kept under for a year with milk and knotgrass."
                    Polygonum aviculare, L.

That of Aubrey in his Nat. Hist. of Wilts, p. 51, was according to Dr. Maton, (Linn. Trans. Vol. v,) and Britton's Beauties of Wiltshire, (v. ii. p. 79),

Agrostis stolonifera, L.

KOHL-RABI, a German name, from It. *cavolo-rapa*, Fr. *chou-rave*, L. *caulo-rapum*, a cabbage-turnip, a cabbage whose stem is swollen so as to resemble a turnip,

Brassica oleracea, L. v. gongylodes, L.

LABURNUM, an adjective from L. *labor*, denoting what belongs to the *hour* of *labour*, and which may allude to its closing its leaflets together at night, and expanding them by day, Cytisus Laburnum, L.

LAD's LOVE, the southernwood, see BOY's LOVE.

LADDER TO HEAVEN, the Solomon's seal, called so, Parkinson tells us, (Th. Bot. p. 699,) "from the forme of the stalke of the leaves, one being set above the other," but more probably from a confusion of *scel* de Notre Dame, our Lady's seal, with *échelle* de N.D. our Lady's ladder. See below LADY's SEAL. Convallaria Polygonatum, L. in Hudson, by mistake, Polemonium cæruleum, L.

LADY's BEDSTRAW, see BEDSTRAW.

Lady in the names of plants almost always alludes to Our Lady, Notre Dame, the Virgin Mary, and often replaces, and is often replaced by that of Venus. Thus Our Lady's comb is the Venus' comb, etc.

LADY's BOWER, so named by Gerarde, p. 740, from "its aptness in making of arbors, bowers, and shadie covertures in gardens," Clematis Vitalba, L.

LADY's COMB, from the long slender parallel beaks of the seed-vessels, Scandix Pecten Veneris, L.

LADY's CUSHION, from its close cushion-like growth, thrift, Armeria vulgaris, W.

9

LADY FERN, from its Latin name in modern works,

Asplenium filix fœmina, Bern.

LADY'S FINGERS from its palmate bracts,

Anthyllis vulneraria, L.

LADY'S GARTERS, from the ribbon-like striped leaves, a variety of Digraphis arundinacea, P.B.

LADY'S HAIR, the quake-grass, Briza media, L.

LADY'S LACES, dodder, from its string-like stems,

Cuscuta.

LADY'S LOOKING-GLASS, from the resemblance of its expanded flower set on the elongated ovary to an ancient metallic mirror on its straight handle,

Campanula hybrida, L.

LADY'S MANTLE, from the shape and vandyked edge of the leaf, a translation of its Arabic name *al kemelyeh*, Sw. *Mariekåpa,* Alchemilla vulgaris, L.

LADY'S NAVEL, see NAVELWORT,

Umbilicus pendulinus, DC.

LADY'S NIGHTCAP, in Wiltshire, (Akerman) the larger bindweed, Convolvulus sepium, L.

LADY'S SEAL, or -SIGNET, M. Lat. *Sigillum Stæ. Mariæ,* a name that in the older writers is correctly given to a Convallaria, the plant now called *Solomon's seal,* from round cicatrices on the root-stock, which resemble the impressions of a seal, but has been injudiciously transferred to a different plant, the black bryony, which has no such characteristic markings. This change seems to have been introduced by the herb-sellers. The latter plant, the bryony, is described by Fernelius and others as one, "quæ *herbariis* et *officinis* sigillum beatæ Mariæ nuncupatur." Cæsalpinus, and Matthiolus tell us that the former, the Convallaria, was called indifferently *St. Mary's* or *Solomon's seal.* "Sunt qui polygonatum *sigillum S. Mariæ,* et qui *sigillum Salomonis* vocitent." Matt. in Diosc. l. iv. c. 5. "It is al one herbe, Solomon's seale, and our Lady's seale," says the Grete

Herbal. See Casp. Bauhin, de Plantis sanctis, p. 67.
The original and right plant,

Convallaria polygonatum, L.

The plant of modern writers,   Tamus communis, L.

LADY'S SLIPPER, from the shape of the labellum of its flower,   Cypripedium Calceolus, L.

LADY'S SMOCK, from the resemblance of its white flowers to little smocks hung out to dry, as they used to be once a year, at that season especially,

> When daisies pied and violets blue,
>   And *lady-smocks all silver white*,
> And cuckoo-buds of yellow hue,
>   Do paint the meadows with delight.
> When shepherds pipe on oaten straws,
>   And *maidens bleach their summer smocks*, etc.
> Shaksp. L.L.L. v. 2.

Cardamine pratensis, L.

LADY'S THIMBLE, called also WITCH'S THIMBLE,

Campanula rotundifolia, L.

LADY'S THISTLE, the milk thistle. See above JUNO'S ROSE.   Carduus Marianus, L.

LADY'S TRESSES, from the resemblance of the flower-spikes, with their protuberant ovaries placed regularly one over the other, to a lady's hair braided,

Neottia spiralis, Rich.

LAKE WEED, from its growth in still water,

Polygonum Hydropiper, L.

LAMB'S LETTUCE, formerly classed with the lettuces, and called in Latin *Lactuca agnina*, " from appearing about the time when lambs are dropped ;" See Martyn, Fl. rustica. or according to Tabernæmontanus, (i. 475), because it is a favourite food of lambs,   Valerianella olitoria, L.

LAMB'S QUARTERS, properly *Lammas quarter*, from its blossoming about the first of August, old style, the day of a festival instituted as a thanksgiving for the first fruits of the harvest, when an oblation was made of loaves of the new corn, the A.S. *hlaf-mæsse*, loaf mass, *missa panum*, in the Salisbury Manual called *Benedictio novorum fructuum*,   Atriplex patula, L.

LAMB'S TOE, from its soft downy heads of flowers,
        Anthyllis vulneraria, L.

LAMB'S TONGUE, Gr. ἀρνογλωσσον, from the shape of the leaf,   Plantago media, L.

LANCASHIRE ASPHODEL, a plant allied to the asphodels and abundant in Lancashire, Narthecium ossifragum, L.

LANG DE BEEF, Fr. *langue de bœuf*, from the tongue-shaped papillated surface of the leaf,
        Helminthia echioides, Gärt.

LARCH, It. *larice*, G. *lärche*, Gr. λαριξ, a name that it may have derived from its use in building and carpentry, L. *lar*, a house, O.N. and Russ. *lar*, a chest, Gr. λαρναξ,
        Pinus Larix, L.

LARK-SPUR, -HEEL, -TOE, or -CLAW, from the projecting nectary,   Delphinium, L.

LAUREL, Sp. *laurel*, L. *laurellus*, dim. of *laurus*, a name originally applied to the sweet bay called in Chaucer, Douglas, and other early writers *laurer*, from Fr. *laurier*, but subsequently extended to other evergreens, and at present in common parlance confined to the cherry laurel,
        Prunus Laurocerasus, L.

„ ALEXANDRIAN-, from Paris, who is called in Homer Alexander, having been crowned with it as victor in the public games, (Stapel in Theophrast. p. 253), whence its names in Apuleius, c. 58, *Daphne Alexandrina*, and *Stephane Alexandrina*,   Ruscus racemosus, L.

„ BAY-,   Laurus nobilis, L.

„ COPSE-, or SPURGE-, from its place of growth,
        Daphne Laureola, L.

„ PORTUGAL-, from its native country,
        Prunus lusitanica, L.

LAUREL, ROMAN-, from its being used in the chaplets worn by the Roman emperors, the sweet bay, Laurus nobilis, L.

LAURESTINE, or more commonly LAURESTINUS, L. *laurus tinus*, from being regarded as a laurel, and as the shrub described by Pliny, and by Ovid (Met. x. 98), under the name of *Tinus*, Viburnum Tinus, L.

LAVENDER, by change of *l* to *r* from Du. and G. *lavendel*, It. *lavandola*, M. Lat. *lavendula*, from *lavare*, wash, as being the plant used to scent newly-washed linen, whence the expression of " laid up in lavender ;" or, as Diez tells us, from its being used at the baths in washing the body. In support of this last opinion C. Stephan tells us (De re hort. p. 54): "*Lavendula* autem dicta quoniam magnum vectigal Genuensibus præbet in Africam eam ferentibus, ubi lavandis corporibus Lybes ea utuntur, nec nisi decocto ejus abluti mane domo egrediuntur." Lavandula Spica, L.

  ,, SEA-, Statice, L.

LAVER, A.S. *læfer*, L. *laver*, a name given by Pliny to some unknown aquatic plant, now applied to certain esculent sea-weeds, as Porphyra laciniata, Ag.

and Ulva latissima, Grev.

LEEK, a remnant of A.S. *porleac*, from L. *porrum*, and *leac*, a plant, G. *lauch*, Du. *look*, Allium Porrum, L.

LENT-LILY, the daffodil, from the season of flowering, the spring, A.S. *lencten*, O.H.G. *lenzo*,

                     Narcissus Pseudonarcissus, L.

LENTILS, Fr. *lentille*, Ervum Lens, L.

LEOPARD'S BANE, Gr. παρδαλιαγχης from παρδαλις, a pard, and ἀγχω, choke, the name of some poisonous plant, which Nicander says in his Theriaca was used on Mount Ida to destroy wild beasts, transferred by Turner to the trulove, a very innoxious one, Paris quadrifolia, L.

LETTUCE, L. *lactuca*, from Gr. γαλα, γαλακτος, milk, and ἐχω, contain, through *lattouce*, an older form of the word that is still retained in Scotland, L. sativa, L.

LETTUCE, FROG'S-,   Potamogeton densus, L.
  ,, LAMB'S-,   Valerianella olitoria, Mn.
  ,, WALL-, a plant of the lettuce tribe found upon
walls,   Prenanthes muralis, L.

LICHEN, Gr. λειχην, a tetter, from its roundish, leprous-looking apothecia, as seen upon old buildings, Lichen, L.

LICHWALE, or, as in a MS. of the fifteenth century, LYTHEWALE, stone-switch, the gromwell, so called in allusion to its stony seeds, and their medicinal use in cases of calculus, from Gr. λιθος, a stone, through M.Lat. *licho* or *lincho*, a pebble, as in the Grant herbier, where the lapis demonis is called *lincho-* and *licho-*demonis, and *wale*, O.Fr. *waule*, now *gaule*, from the Breton *gwalen*, a switch,
                                  Lithospermum officinale, L.

LICHWORT, from its growing on stones (see LICHWALE), the wall-pellitory,   Parietaria officinalis, L.

LILAC, a Persian word introduced with the shrub,
                                  Syringa vulgaris, L.

LILY, L. *lilium*, Gr. λειριον, of unknown, very ancient origin, used in some oriental languages for a flower in general, as in Cant. vi. 2-3, and Mat. vi. 28, and as *rosje*, rose, is used in the Illyrian; a trope of frequent occurrence among all nations, particularly the less cultivated races.
                                  Lilium.

  ,, CHECKERED-, the fritillary, from the markings on its petals,   Fritillaria Meleagris, L.
  ,, WATER-,   Nymphæa alba, L.

LILY-AMONG-THORNS, of Canticles ii. 2, L. *Lilium inter spinas*, understood by the herbalists as the woodbine, which, as W. Bulleyn says, "spredeth forth his sweete lilies like ladies' fingers among the thorns,"
                                  Lonicera Caprifolium, L.

LILY-OF-THE-VALLEY, or LILY-CONVALLY, L. *lilium convallium*, lily of combes or hollows, a name taken from Cant. ii. 1, "I am the lily of the valleys,"
                                  Convallaria majalis, L.

LIME, LINE, or LINDEN-TREE, called in all Germanic languages, and in Chaucer, *Linde*, a word connected with Ic. and Sw. *linda*, a band, and A.S. *līðe*, pliant, which stands in the same relation to the continental name, as, e.g. *hriðer*, cattle, to G. *rind*, and *toð* to Fris. *tond*, that is having a final *d* changed to *ð*, and the *n* omitted. The name has evidently been originally applied to the inner bark, or bast, of the tree so much used in the North for cordage. In the Herbals, and all old works after Chaucer's time, it is spelt *Lyne* or *Line*, as in the ballad of Robin Hood and Guy of Gisborne, where it rhymes to "thine,"

> "Now tell me thy name, good fellow," said he,
> "Under the leaves of *lyne*."

The *n* has in later writers been changed to *m*, and *lyne* become *lime*, as *hollen* hol*m*, *henep* hem*p*, and *mayne* maim. *Linden* is the adjectival form of *lind* with 'tree' or 'timber' understood, and it is to be remarked that the names of most trees are properly adjectives, and in the Western counties are generally used with an adjectival termination, as elmen-tree, holmen-tree, &c.   Tilia europæa, L.

LINE and LINSEED, L. *linum*, Gr. λινον, flax, probably a word adopted from a language alien to the Greek, upon the introduction of its culture,   Linum usitatissimum, L.

LING, Da. Nor. and Sw. *lyng*, a word which Holmboe considers to represent Skr. *gangala*, by a replacing of *g* with *l*, the common heath, possibly a form of A.S. *lig*, fire, as implying "fuel," and connected with L. *lignum*, firewood. This word is often combined with *hede*, a heath, as in Sw. *ljunghed*, Da. *lynghede*, ericetum, a heath-land, and conversely *hedelyng*, the heath-plant; leading to the belief that *heath* was the waste land, and *lyng* the shrub growing on it. See Diefenbach (Lex. Comp. ii. 496.)

Calluna vulgaris, L.

LION'S-FOOT, or -PAW, from the shape of the leaf resembling the impress of his foot, Alchemilla vulgaris, L.

LIQUORICE-VETCH, a vetch-like plant with a sweet root, M. Lat. *liquiricia*, from L. *glycyrrhiza*, Gr. γλυκυς, sweet, and ῥίζα, root,   Astragalus glycyphyllus, L.

LIRY-CONFANCY, a corruption of L. *lilium convallium*, lily of the valleys,   Convallaria majalis, L.

LITHY-TREE, from A.S. *liðÞ*, pliant, a word etymologically identical with *lind* (See LINDEN,); the tree being so called, because, as Parkinson says: (Th. Bot. p. 1448,) "the branches hereof are so tough and strong withall, that they serve better for bands to tye bundels or any other thing withall, or to make wreathes to hold together the gates of fields, then either withy or any other the like," the wayfarer tree,   Viburnum Lantana, L.

LITMUS, G. *lackmus*, from *lac*, Skr. *laksha*, a red dye, and *moos*, moss, a lichen, in popular language a moss, used in dyeing,   Roccella tinctoria, DC.

LITTLEGOOD, a plant so called on the Eastern Border (Johnst.) to distinguish it from the allgood,

          Euphorbia Helioscopia, L.

LIVELONG, or LIBLONG, from its remaining alive hung up in a room. Brande in Pop. Ant. says that it is a habit with girls to set up two plants of it, one for themselves and another for their lover, upon a slate or trencher, on Midsummer eve, and to estimate the lover's fidelity by his plant living and turning to theirs, or not. The name should probably be "Livelong *and* Liblong" (Live long and Love long). See MIDSUMMER MEN.   Sedum Telephium, L.

LIVERWORT, from the liver shape of the thallus, and its supposed effects in disease of the liver. See Brunschwygk, (b. ii. c. 11).   Marchantia polymorpha, L.

,, GROUND-,   Peltidea canina, Ach.

,, NOBLE-, in America called *Liverleaf*, and from its three-lobed leaves supposed to be, as Lyte tells us, (b. i. ch. 40,) "a sovereign medicine against the heate and inflammation of the liver,"   Anemone hepatica, L.

LOBGRASS. from *lob,* or *lop,* to loll or hang about, as in *loblolly,* etc., so called from its hanging panicles,

Bromus mollis, L.

LOCKEN, or LUCKEN GOWAN, or GOWLON, a closed goole or goldin, a term applied, according to Lightfoot and Jamieson, to the globe flower, called for the same reason, viz., its connivent petals, the cabbage daisy,

Trollius europæus, L.

LOGGERHEADS, from the resemblance of its knobbed involucres to a weapon so called, consisting of a ball of iron at the end of a stick, the knapweed, the Clobbewed of old MSS.,

Centaurea nigra, L.

LONDON PRIDE, a name given in the first place to a speckled Sweet William, from its being a plant of which London might be proud, and similar to that of the Mountain Pride, the Pride of India, and the Pride of Barbadoes, (see Parkinson's Parad. p. 320,) but of late years transferred to a saxifrage, which is commonly supposed to be so called, because it is one of the few flowers that will grow in the dingy lanes of a town. See Seeman's Journal, vol. i. It is understood, however, upon apparently good authority, that of Mr. R. Heward in the Gardener's Chronicle, to have been given to this latter plant in reference to the person who introduced it into cultivation, Mr. London, of the firm of London and Wise, the celebrated Royal Gardeners of the early part of the last century.   Saxifraga umbrosa, L.

LONDON ROCKET, called *rocket* from its leaves resembling those of an eruca, and *London* from its springing up abundantly in 1667 among the ruins left by the Great Fire,

Sisymbrium Irio, L.

LONG PURPLES, of Shakspeare's Hamlet, (iv. 7,) supposed to be the purple-flowered orchis,   O. mascula, L.

LOOSESTRIFE, a translation of the Lat. *lysimachia,* as though the plant were called so from its stopping strife, Gr. λυσι and μαχη.   Pliny tells us (b. xxv. c. 35) that

the name was given to it after a certain king Lysimachus ;
but, nevertheless, in deference to a popular notion, he adds
that, if it be laid on the yoke of oxen, when they are
quarrelling, it will quiet them.   Lysimachia vulgaris, L.

LOOSESTRIFE, PURPLE-,   Lythrum Salicaria, L.

LORDS AND LADIES, from children so calling the spadix
of the Wake Robin, as they find it to be purple or white ; a
name of recent introduction, to replace certain older, and
generally very indecent ones ;   Arum maculatum, L.

LORER, Fr. *laurier*, the bay tree, in Chaucer and Gower's
works,   Laurus nobilis, L.

LOUSE-BERRY TREE, from its fruit having once been used
to destroy lice in children's heads : " The powder kills nits,
and is good for scurfy heads."   Dict. of Husbandry, under
" Spindle tree ;" and Loudon, (Arb. Brit. ii. 406) ;
                         Evonymus europæus, L.

LOUSE-BUR, from its burs, or seed-pods, clinging like
lice to the clothes,   Xanthium strumarium, L.

LOUSE-WORT, "because," says Gerarde, p. 913, "it filleth
sheep and other cattle, that feed in meadows where this
groweth, full of lice,"   Pedicularis, L.

LOVAGE, in Pr. Pm. and in Holland's translation of Pliny,
spelt *Love-ache*, as though it were love-parsley, Fr. *levesche*,
corruptions of Lat. *levisticum*, whence also, through the
same mistake, G. *liebstöckel*, and A.S. *lufestice* and *lube-
stice*,   Levisticum officinale, Ko.

LOVE, the virgin's bower : " The gentlewomen call it
Love," says Parkinson, (Th. Bot. p. 384), from its habit of
embracing, perhaps,   Clematis Vitalba, L.

LOVE-APPLES, L. *poma amoris*, Fr. *pommes d'amour*, from
It. *pomi dei Mori*, Moors' apples, this fruit having been introduced as *mala æthiopica*, Solanum Lycopersicum, L.

LOVE-IN-A-MIST, or -IN-A-PUZZLE, from its flower being
enveloped in a dense entanglement of finely divided bracts,
                         Nigella damascena, L.

LOVE-IN-IDLENESS, or LOVE-AND-IDLE, or, with more accuracy, LOVE-IN-IDLE, *i.e.*, in vain, as in the phrase in Exod. xx., 7 : A.S., " Ne nem þu Drihtnes namen *on ydel"* —" Tac þu noght *in idel* min namen," a name of the pansy that perpetuates a current phrase, as in the couplet,

" When passions are let loose without a bridle,
Then precious time is turned to *love and idle ;*

Taylor.

but why it was attached to this flower, is not apparent.

Viola tricolor, L.

LOVE-LIES-BLEEDING, from the resemblance of its crimson flower-spike to a stream of blood, and the confusion of the two first syllables *Amar* of its Latin name with *amor,* love, Amaranthus caudatus, L.

LOVEMAN, the goosegrass, a name given to it by Turner to express the Gr. φιλανθρωπος, from its clinging to people, Galium Aparine, L.

LOWRY, L. *laurea,* adj. of *laurus,* laurel, the spurge laurel, so called from its evergreen leaves, Daphne Laureola, L.

LUCERNE, apparently from the Swiss canton of that name, but Diez says that its derivation is unknown. By some of the older herbalists the sainfoin was called so. At present the name is confined to the Medicago sativa, L.

LUJULA, contracted from It. *Alleluiola,* dim. of *Alleluia ;* see HALLELUJAH.

LUNARIE, L. *lunaria,* from *luna,* moon, a name given to a great number of different plants. The following description of one is copied by Gesner, in his treatise upon plants called Lunaria, from some anonymous author : " *Lunaria* emicat in montibus humidis. Caule visitur procero et rubente, anguloso, nigris asperso maculis. Folia instar Lunæ orbis, aut nummi alicujus, rotunda sunt, aut sampsuchi foliis similia. Flos luteus ; odor moschi. Chymistæ facultatem fere eandem, quam chelidonio ei attribuunt. Metalla omnia vi ejus in Solem et Lunam converti pollicen-

tur. Noctibus lucere aiunt, crescente Luna. Colligi
oportere æstate pridie Divi Joannis Baptistæ, Luna plena,
ante ortum Solis ; crescit enim (aiunt), et decrescit cum
Luna." Gesner observes that "the chymists, who first
wrote in this way, and those who have since believed them,
were either ignorant men, who easily and rashly placed
reliance in the fables of others, and published monstrous
descriptions of plants that have no existence ; or signified
by enigmatical descriptions, not any plants, but things very
different, as by the name of *chelidonium* and *elydrium*, I
know not what quintessence or philosopher's stone." As at
present understood, it is the fern that from its semilunar
fronds is called Moonwort,   Botrychium Lunaria, L.

Lung-flower a translation of Gr. πνευμονανθη, from
πνευμων, lungs, and ἀνθος, flower,
                         Gentiana Pneumonanthe, L.

Lung-wort, L. *pulmonaria*, from *pulmo*, lungs, being
supposed, from its spotted leaves, to be a remedy for diseased lungs,   P. officinalis, L.

   „ Tree-,   Sticta pulmonaria, Hook.

Lupine, L. *lupinus*, literally "wolfish," but supposed to
be related to Gr. λοπος or λοβος, a husk, and λεπω, hull or
peel.   If we had any reason to think it an Asiatic genus
that was introduced into Italy, we might explain its name
as a translation of λυκειος, Lycian, mistaken for an adjective
from λυκος, a wolf.   Wedgwood's derivation of it from a
Slavonian root is, for reasons geographical and historical,
quite inadmissible.   Lupinus, L.

Lustwort, a name translated from Du. *loopich-cruydt*,
which, according to Dodoens, has that meaning, and has
been given to the plant, he says, "quia acrimonia sua
sopitum Veneris desiderium excitet," the sundew,
                                  Drosera, L.

Lyme-grass, from L. *elymus*,   E. europæus, L.

Lyon's snap, from L. *Leontostomium*, Gr. λεοντος, and

στομιον, the snapdragon, *snap* having formerly had the
meaning of the Gr. στομιον, a little mouth,

Antirrhinum majus, L.

MADDER, in old MSS. *madyr*, from the plural of L. Ger.
made, a worm, and called in an Anglo-Saxon MS. of the
thirteenth century *vermiculum*. See Mayer and Wright,
p. 139. *Made* is the same word as the Go. and AS. *maδa*,
whence *mad*, used by Tusser for a maggot, and *moth*, which
properly means the worm "that fretteth the garment,"
and not its winged imago, a word related to Go. *matjan*, eat,
L. *mandere*, its root *mad*. The name was applied to the
plant now called so from confusion with another red dye,
that was the product of worms, viz., the *cocci ilicis*, which
infest the Quercus coccifera, L., and which were called in
the middle ages *vermiculi*, whence Fr. *vermeil* and *vermillon*,
a term now transferred to a mineral colour.

Rubia tinctorum, W.

,, FIELD-,   Sherardia arvensis, L.

MADNEP, the mead-nape, or -parsnep, or as it was once
spelt, pas-nep, the cowparsnep. From Gerarde's assertion
that "if a phrenetiche or melancholic man's head be anointed
with oyle wherein the leaves and roots have been sodden,
it helpeth him very much," it would seem as though *pas-nep*
was misunderstood as It. *pazzo-napo*, mad turnep, and *mead*
conformably changed to *mad*.

Heracleum Sphondylium, L.

MAD-WORT, Du. *meed*, madder, for which its root was
used,   Asperugo procumbens, L.

MAGHET, maid, an adopted Flemish word, Go. *magaþs*,
O.H.G. *magad*, Fris. *mageth*, A.S. *mægδ*, Fl. *maghet*, as in
the hymn beginning

"O, moeder ende maghet, reine vrouwe!"

Willems, No. 114.

Lat. *flos virgineus*, Gr. παρθενιον, a name given to many

radiate compositæ with white ray florets, says Stapel in
Theophrast. (p. 833, b.) "quod morbis mulierum uter-
inis medeantur ;" an idea suggested by their fancied
resemblance to the moon, which, from its regulating the
monthly periods of the year, was supposed to influence
the complaints peculiar to young women, and all affections
of the womb. See MAITHES, MARGUERITE, MATHER,
MAUDLIN, MAYDWEED, MAYWEED, and MOONWORT. These
plants were, in ancient times, and for the same reason,
dedicated to the virgin goddess of the night, the θεος των
καθαρματων, or Diana; but in Christian times have been
transferred to the two saints, who in this particular replace
her, St. Mary Magdalene and St. Margaret.

Anthemis, Achillæa, Pyrethrum, Bellis, Chrysanthemum,
Matricaria, &c.

MAIDEN-HAIR, from its hair-like fine stalks,

Adiantum Capillus Veneris, L.

According to Lobel (Kruydtboek, p. 126), the name of
Mayden heere was in his time given to the bog asphodel,
"om dat de jonge dochters haer hayr daermede geel
maecken ;" because young girls make their hair yellow
with it. This fashion of dyeing the hair yellow was very
prevalent in the middle ages, but the lye of wood ashes
was most generally used for the purpose, and as the em-
ployment of this plant is not noticed by other writers, nor
any such name given to it, he was probably mistaken. See
MAID's HAIR.   Narthecium ossifragum, Lam.

MAIDEN-HAIR-GRASS, G. (in Bauhin's Th. Bot.) *jungfrauen
haar*, but by Parkinson (Th. Bot. p. 1166,) spelt *Mead Hair-
grass;* in either case from its delicate hair-like stalks ; the
quaking-grass,   Briza media, L.

MAIDEN-PINK, a mistake for MEAD-PINK, G. *wiesen-nelke*,
a pink that grows in meadows,   Dianthus deltoides, L.

MAID's HAIR from its soft flocculent habit, like the loose
un-snooded hair of maidens, and its yellow colour, to

which, as a beauty in the hair of women, such frequent allusion is made by Chaucer and other romance writers. Even so late as Henry the Eighth's reign Horman says, "Maydens were silken callis, with the whiche they keepe in ordre theyr heare made yelowe with lye." See Way's Promp. Pm. p. 294.   Galium verum, L.

MAITHES, that is *maids*, A.S. *mageƍe, magƍe*, in Pr. Pm. and other old works *mayde-wede* and *maythys*, from Gr. παρθενιον, because, says W. Coles, "it is effectual against those distempers of the womb to which virgins are subject," meaning hysterics, and other irregularities of the system. See MAGHET.   Pyrethrum Parthenium, L.

   „   RED-, or RED MAYDE-WEED, from its having been classed with the composite flowers called *maithes*, and its crimson colour,   Adonis autumnalis, L.

MAKEBATE, because, says Skinner, "if it is put into the bed of a married couple, it sets them quarrelling," but a mere translation of its Latin name as if from πολεμος, war,
                  Polemonium cæruleum, L.

MALLOW, A.S. *malwe*, L. *malva*, Gr. μαλαχη, which Pliny and Isidore derive from μαλασσειν, soften, as alluding to the laxative property of the plant,   Malva.

   „   MARSH-,   Althæa officinalis, L.
   „   MUSK-,   M. moschata, L.
   „   TREE-,   Lavatera arborea, L.

MANDRAKE, Gr. μανδραγορας, a plant generally believed to have been one nearly related to the deadly night-shade. See Hogg in Hooker's Journal, 2nd ser. vol. i. p. 132. Fraudulent dealers used to replace its roots with those of the white bryony cut to the shape of men and women, and dried in a hot sand bath. See Brown's Popular Errors, b. ii. ch. 6 ; Tragus, ch. cxxvi. ; Stapel in Theophrast. p. 583 ; and especially Matthioli, l. iv. c. 61, where he tells us that Italian ladies in his own time had been known to pay as much as 25 and 30 ducats [aureos] for one of the artificial

Mandrakes of certain itinerant quacks; and describes the process by which they are made. They were supposed to remove sterility, a notion that has prevailed in the East from the remotest antiquity. Hence Rachel's desire to obtain them, as related in Gen. ch. xxx., v. 14. That of these dealers was the common white bryony,

Bryonia dioica, L.

MANGEL WURZEL, literally "scarcity root," but originally *Mangold*, a word of unknown meaning, and as *Mengel* or *Menwel* applied to docks,

Beta vulgaris, L. var. hybrida, Sal.

MANNA-GRASS, from the sweet taste of the seed,

Glyceria fluitans, RB.

MAP-LICHEN, from the curious map-like figures formed by its thallus on flat stones, Lecidea geographica, Hook.

MAPEL, A.S. *mapel-treow*, or *mapulder*, in Pr. Pm. *mapulle*, a word adopted from the language of the ancient Britons, and of early and general use throughout England, as is shown by the number of places named after the tree. It is clearly the Welsh *mapwl*, a knob in the middle of any thing, and refers to the knotty excrescence from the trunk of the tree, the *bruscum*, so much employed and so highly valued in the Roman times and in the Middle ages for making bowls and tables, that single specimens of it have fetched many thousand pounds. See Pliny, N.H. xvi. 27, and Smith, Dict. Ant. art. Mensa. The tree was naturally named after its most valuable product, its *mapwl*, and the word adopted, upon their arrival here, by the Anglo-Saxons.

Acer campestre, L.

MARAM, either the Gael *muram*, or, as is more probable, the Fris. and Dan. *marhalm*, sea-haulm or straw, a word which in Norway is applied to the zostera and certain fuci, the grass called mat-weed, Psamma arenaria, PB.

MARCH or MERCH, Da. *märke*, Sw. *mærki*, the old name of parsley, preserved in *stanmarch*, the Alexander, and in the G. *wasser-merke* and Da. *vand-merke,* celery, formed from L. *armoracia,* in the fifteenth century called in German *merich* and *mirrich,* and to this day in Wetterau *mirch.* In A.S. the parsley is called *merce, meric,* and *merici.* See L. Diefenbach, Or. Eur. No. 26. Apium, L.

MARE-BLOBS, from A.S. *mere* and *myre,* a marsh, and also a mare, and *blob* or *bleb,* a bladder, the marsh-mallow, so called from its round flower-buds,

Caltha palustris, L.

MARE'S-TAIL, a plant called in old herbals "Female Horse-tail," Lat. *cauda equina fœmina,* being looked upon as the female of the larger and stronger Equisetum fluviatile. Modern botanists, following Hudson, have shifted the hyphen, and chosen to understand the name as "Female-horse Tail," or "Mare's Tail." Hippuris vulgaris, L.

MARGUERITE, from the French, the daisy, in Chaucer MARGARETTE, a plant so called, not from its fancied innocence and simplicity, but because, for reasons given under MAGHET, it was formerly used in uterine diseases, which were under the especial care of *St. Margaret,* of Cortona. This lady, according to Mrs. Jameson, (Mon. Ord. p. 329), had for some years led an abandoned life, but had repented and been canonised, and was regarded by the people of her native town as a local Magdalene, and, like her prototype, supposed, in respect of her early habits, to preside over the diseases of the womb, and others peculiar to young women. See MAUDLIN. Her name has been mixed up with that of a *St. Margaret,* of Antioch, who, according to Hampson, (Med. Æv. Kal. ii. 257), "was invoked as another Lucina, because in her martyrdom she prayed for lying-in-women." But it is the Cortona saint, and not this one of Antioch, whose name has been given to the daisy, and probably in the first place to the moon-daisy. The story of the maiden of Antioch,

"Maid Margarete, that was so meeke and mild;"

10

to whom it is popularly assigned, is given by Mrs.
Jameson, in Sacr. and Legend. Art. p. 306.

Bellis perennis, L.

MARIET, the Coventry bell, its French name, L. *viola
Mariana,*   Campanula Trachelium, L.

MARIGOLD, called in the Grete Herball *Mary Gowles*, a
name that seems to have originated from the A.S. *mersc-
mear-gealla*, marsh-horse-gowl, the marsh marigold, or
caltha, transferred to the exotic plant of our gardens, and
misunderstood as "Mary gold." Its foreign synonyms
have no reference to the Virgin Mary. It is often mentioned by the older poets under the name of *Gold* simply.

Calendula officinalis, L.

  ,,   CORN-, from its place of growth, and yellow flowers,
Chrysanthemum segetum, L.

  ,,   MARSH-, A.S. *mersc-mear-gealla*, the second term
of which, meaning properly "horse," has been understood
as "Mary." See MARE-BLOBS, and GOOL.

Caltha palustris, L.

MARJORAM, or MAJORAM, L. *majorana*, with change of
n to *m*, as in Lime, Holm, etc. Origanum Majorana, L.

MARMARITIN, in Middleton's "Witch," Lat. in Pliny
marmaritis, the peony,   Pæonia corallina, L.

MARROW, VEGETABLE-, from its soft and delicate flesh,
Cucurbita ovifera, W.

MARSH-ASPHODEL, a plant of the asphodel tribe growing on moors,   Narthecium ossifragum, L.

MARSH-BEETLE, or -PESTLE, from its shape, the reed-
mace,   Typha latifolia, L.

MARSH-CISTUS,   Ledum palustre, L.
MARSH-ELDER,   Viburnum Opulus, L.
MARSH-FERN,   Aspidium Thelypteris, Sw.
MARSH-FLOWER,   Limnanthemum nymphæoides, Lk.
MARSH-HOLYROSE,   Andromeda polifolia, L.
MARSH-MALLOW,   Althæa officinalis, L.

MARSH-MARIGOLD,   Caltha palustris, L.
MARSH-PARSLEY,   Apium graveolens, L.
MARSH-PENNYWORT,   Hydrocotyle vulgaris, L.
MARSH-SAMPHIRE,   Salicornia herbacea, L.
MARSH-TREFOIL,   Menyanthes trifoliata, L.
MARSH-WORTS,   Vaccinium Oxycoccos, L.
MARY-BUD, in the aubade in Cymbeline, (A. ii. sc. 3),

> "And winking *Marybuds* begin
> To ope their golden eyes:"

the marigold, Columella's "flaventia lumina calthæ:" l. 97:

> "The Marigold that goes to bed with the sun,
> And with him rises weeping:"—Wint. T. A. iv. sc. 3 :

in allusion to its flowers, which, as Lyte says, "do close at the setting downe of the sunne, and do spread and open againe at the sunne rising:" a phenomenon to which the older poets allude with great delight, both in respect to this flower and the daisy.   Calendula officinalis, L.

MASER-TREE, the maple, from the bowls or drinking cups, called *masers*, being made from the knotty parts of its wood, called in O.H.G. *masar*, whence M. Lat. *scyphi maserini*, Du. *maes-hout*, from *maese*, a spot, O.H.G. *mazeldera*, masel-tree, whence G. *massholder*. See MAPLE. Bowls made of silver and gold were called by the same name, as in Ritson's Ancient Popular Poetry, p. 77,

> "Pecys of syluyr, *masers* of golde."

See Pr. Pm. p. 328, Way's note.   Acer campestre, L.

MASTER-WORT, a translation of its Latin name, *Imperatoria*, which was understood by the herbalists as indicating the masterly virtues of the plant, but has more probably been given to it, with the sense of "Imperial," under the idea that it was the one described by Apuleius, c. 130, as having been used by Augustus under the names of *Basilica* and *Regia*, as a protection against serpents: "Hac utebatur *Imperator* Augustus."   Imperatoria Ostruthium, L.

MATFELLON, from L. *maratriphyllon*, fennel-leaf, Gr.

μαραθρου φυλλον, called in the Ort. Sanit. c. 432, and in
the Grete Herball, *Marefolon*, in Gerarde *Matfellon*, in
Dodoens *Materfillon* and "*Matrefilon*, voce, ut apparet,
corrupta," in W. Coles *Madefelon*, in Parkinson "*Matre-
fillon*," in old MSS. *Mattefelone, Maudefelune, Madfeloun*,
etc. The Lat. *maratriphyllon*, the source of all these barbarous terms, seems in the first place to have been given
to the water violet, Hottonia, on account of its finely
divided fennel-like leaves, and this is the plant which
bears the name in Lobel and Pena's Kruydtboek, 1581,
p. 965. From this it would seem to have been extended
to other so-called violets, viz. the genus Viola, and the
centauries. Thus in the Grete Herball, cap. ccccii, we
read "Jacea, Herba clavellata, Torquea, Marefolon. Jacea
is an herbe," etc. In H. Brunschwygk, p. xlix, these
synonyms are assigned to the pansy : " Freissam krut von
den kriechen torqueta, und von den arabischen marefolon
genant, und in latin yacea oder herba clavellata, ouch von
ettichen dreifaltigkeit blumen genant umb dreyerley farb
siner blůmen, gelb, blow, und weyss." In the Ortus Sanitatis also it is figured and described as a pansy, under the
German name Freyschem-kraut, epilepsy-wort, and en-
titled "Jacea vel herba clavelata, Latine, grece torqueta,
arabice marefolon." *Jacea* being extended to the genus
Centaurea has carried the name from the violet tribe to
the knapweed, but under the corrupt form of *Matfellon*.

Centaurea nigra, L.

Mat-grass, or Small Mat-weed, from its dense mat-
like tufts,   Nardus stricta, L.

Mat-weed, from its use in making mats,

Psamma arenaria, R.S.

and also the cord-grass, because as Gerarde says, p. 39,
" these kindes of grassie or rather rushie reede serve for to
make mats and hangings for chambers, frailes, baskets,
and such like, and the people of the country where they

grow do make beds of them, and strawe their houses and chambers with them, insteede of rushes."

Spartina stricta, L.

MATHER or MAUTHER, a word used in the Eastern counties to mean a girl of the working class, O.N. *maðr*, a man, a human being, a word from which it is probable that some prefix has been lost, a name applied to the wild chamomiles as a translation of their Greek name παρθενιον. See MAGHET and MAITHE.

Anthemis Cotula, L.

MAUDLIN, MAUDELINE, or MAWDELEYN, L. *herba divæ Mariæ*, It. *herba di santa Maria*, so called after St. Mary Magdalene, either in allusion to her box of scented ointment, as containing this aromatic; or that, like other plants with white ray florets, it was employed in uterine diseases, over which, as the especial patroness of loose women, she was supposed to preside. See COSTMARY and MAGHET.

Balsamita vulgaris, L.

and

Achillæa Ageratum, L.

MAUDLIN-WORT, from its use in the same complaints as the above, the moon-daisy,

Chrysanthemum Leucanthemum, L.

It is necessary to observe that the monks in the middle ages mixed up with the story of the Magdalene, as recorded in Scripture, that of another St. Mary, whose early life was passed in a course of debauchery:

"Seint Marie Egipciake in Egipt was ibore.
Al hire yong lif heo ladde in sinne and in hore,
Unnethe zhe was tuelf ycr old, ar zhe gon do folie,
Hire bodi and al here wille heo tok to sinne of lecherie."

Cott. MS. Julius, D. ix., fol. 52, b, quoted
by Hampson, ii. 257.

Her penance and pardon were a favourite subject for the legends of all Western Europe. The attributes of the impure goddess of the Egyptians, Isis, and of the Greek Artemis, and the Roman Juno Lucina, have been transferred in Roman Catholic times to this saint, and her counterpart, St. Margaret.

MAULE, the mallow, It. and Sp. *maula*, by transposition of the *u*, from L. *malva*,   M. sylvestris, L.

MAWSEED, G. *magsamen*, Pol. *mak*, poppy-seed,

Papaver somniferum, L.

MAWTHER see MATHER.

MAY or MAY-BUSH, from its time of flowering, the hawthorn,   Cratægus Oxyacantha, L.

MAY-LILY, the lily of the valley,

Convallaria majalis, L.

MAYWEED, from *may*, a maiden, Da. *mó*, Ic. *mey*, and not from the month, a plant used for the complaints of young women, Gr. παρθενιον.   See MAGHET.

Pyrethrum Parthenium, L.

„   STINKING-,   Matricaria Chamomilla, L.

MAYDWEED, says Lupton, (b, 8, n. 46), "is a stinking hearbe, having a flower like a dayseye," one so named from being used in the same cases as the mathers and mayweeds,   Anthemis Cotula, L.

MAY-WORT from the month of its flowering,

Galium cruciatum, L.

MAZZARDS, from L. *manzar*, explained in Pr. Pm. by "spurius, pelignus," a wild, a spurious cherry,

Prunus avium, L.

MEADSWEET, see MEADOW SWEET.

MEADOW BOUTS, Fr. *bouton d' or*, the wild bachelor button of moist grass-lands,   Caltha palustris, L.

MEADOW CLARY,   Salvia pratensis, L.

MEADOW CRESS,   Cardamine pratensis, L.

MEADOW PARSNIP,   Heracleum Sphondylium, L.

MEADOW PINK, the ragged Robin,

Lychnis flos cuculi, L.

and the maiden pink,   Dianthus deltoides, L.

MEADOW RUE, from its finely divided rue-like leaves,

whence its name in old writers Peganon or Pigamon, from
Gr. πηγανον, rue, Thalictrum flavum, L.

MEADOW SAFFRON, from the resemblance of its flowers
to those of the crocus or true saffron,
Colchicum autumnale, L.

MEADOW SAXIFRAGE, from its leaves resembling those of
the burnet saxifrage, Silaus pratensis, Bess.

MEADOW SWEET, an ungrammatical and ridiculous name,
a corruption of *mead-wort*, A.S. *mede-* or *medo-wyrt*, Da.
miöd-urt, Sw. *miöd-ört*, the *mead-*, or honey-wine- herb.
Hill tells us in his Herbal, p. 23, that "the flowers mixed
with mead give it the flavour of the Greek wines," and this
is unquestionably the source of the word. Nemnich also
says that it gives beer, and various wines, and other drinks
an agreeable flavour. The Latin *Regina prati*, meadow's
queen, seems to have misled our herbalists to form the
strangely compounded name now in use.
Spiræa ulmaria, L.

MEAD-WORT, or MEDE-WORT, the old and correct name
of the so-called MEADOW SWEET. See above.

MEAL-BERRY, Da. *meelbær*, Norw. *miölbær*, from the
floury character of the cellular structure of its fruit,
Arbutus Uva ursi, L.

MEALY-TREE, from the mealy surface of the young shoots
and leaves, Viburnum Lantana, L.

MEDICK, L. *Medica*, Gr. μηδικη βοτανη, (Diosc. ii., 177),
the name of some plant that according to Pliny (l. xviii.
c. 43) was introduced into Greece by the army of Darius,
and called so to mean *Median*. It seems formerly to have
been given to a sainfoin, but is at present assigned to the
lucerne and its congeners. Medicago sativa, L., etc.

MEDLAR, called in Normandy and Anjou *meslier*, from
L. *mespilus*, but as the verb *mesler* became in English
meddle, so this fruit also, although a word of different
origin, took a *d* for an *s*, and became *medlar*.
Mespilus germanica, L.

MELANCHOLY GENTLEMAN, from its sad colour,
Hesperis tristis, L.

MELANCHOLY THISTLE, from its supposed virtue in the cure of melancholy,   Carduus heterophyllus, L.

MELILOT, L. *melilotus*, Gr. μελιλωτος, from μελι, honey, and λωτος, a name applied by Greek writers to some very different plant from the one at present called so,
Melilotus officinalis, L.

MELON, a word derived in the dictionaries from L. *melopepo*, but more probably an augmentative of its first two syllables *melo*, from Gr. μηλον, an apple, Fr. and Sp. *melon*, M.Lat. *melo*,   Cucumis Melo, L.

MERCHE, see MARCHE.

MERCURY, a name rather vaguely applied in old works, and now limited to a poisonous weed, from the god Mercury, in respect of some fancied activity in its operation; or, according to Pliny, from its having been discovered by him;   Mercurialis, L.

  „ DOG'S-, from its worthlessness,
Mercurialis perennis, L.

  „ ENGLISH-, the all-good,
Chenopodium Bonus Henricus, L.

  „ FRENCH-,   Mercurialis annua, L.

MERCURY'S VIOLET, the Mariet,
Campanula Trachelium, L.

MERRY, Fr. *merise*, mistaken for a plural noun, as cherry from cerise, L. *mericea*, adj. of *merica*, some unknown berry mentioned by Pliny, the wild cherry,
Prunus avium, L.

MEW, Gr. μηον,   Meum athamanticum, L.

MEZEREON, a name derived by Mesue from a Persian name signifying a "destroyer of life," but by others from Arab. *maçzeroun*, a dwarf olive, a name adopted by the herbalists from the writings of the Arabian physicians,
Daphne Mezereon, L.

MICHAELMAS DAISY, from its resemblance to a daisy, and
its season of flowering,   Aster Tradescanti, L.

MIDSUMMER DAISY,
        Chrysanthemum Leucanthemum L.

MIDSUMMER MEN, from a custom of girls to try their
lovers' fidelity with it on Midsummer eve. See LIVELONG.
        Sedum Telephium, L.

MIGNONETTE, dim. of Fr. *mignon*, darling, from G.
minne, love, a name applied in France to several very
different plants,   Reseda odorata, L.

MILDEW, A.S., *mele-deau*, from *melu*, meal, and *deau*,
dew, G. *mehlthau*, a name descriptive of the powdery appearance upon leaves and stems of plants from the growth
of certain minute fungi,   Erisiphe, DC., etc.

MILFOIL, Fr. *mille* and *feuilles*, L. *mille foliola*, from the
numerous fine segments of its leaves, a name under which
Apuleius seems to have meant the horsetail, Equisetum,
describing it (in c. 89) as a plant " thyrso unius radicis,
molli, fulvo, ita coæquato atque elimato, ut manufactus
videatur, foliis fœniculi similibus;" but at present given
to the yarrow,   Achillæa Millefolium, L.

  ,, HOODED-,   Utricularia, L.

  ,, WATER-,   Myriophyllum, L.

and also   Hottonia palustris, L.

MILK PARSLEY, from its milky juice,
        Peucedanum palustre, Mn.

MILK-THISTLE, a thistle supposed to have derived the
colour of its leaves from the milk of the Virgin Mary
having fallen upon them, as she nursed the infant Jesus,
a fable suggested by the similar one of the lily having been
whitened by the milk of Juno as she nursed the infant
Hercules. See JUNO'S ROSE.
        Silybum Marianum, DC.

MILK-VETCH, from a belief that it increased the secretion
of milk in the goats that fed on it,   Astragalus, L.

MILK-WORT, from its "virtues in procuring milk in the breasts of nurses," says Gerarde, p. 450,

Polygala vulgaris, L.

„ SEA-, Glaux maritima, L.

MILL-MOUNTAIN, from the Lat. cha-*mœl*-inum *montanum*, Gr. χαμαι-λινον, ground flax,

Linum catharticum, L.

MILLET, Fr. *millet*, It. *miglietto*, dim. of *miglio*, from L. *milium*, a name which, for want of good distinctive terms, is popularly extended to several different species of the genera Milium, Panicum, Paspalum, and Sorghum.

MILTWASTE, the finger-fern or ceterach of which Du Bartas says, p. 79 (Sylvester's transl. 1611),

"The Finger-ferne, which being given to swine,
It makes their Milt to melt away in fine;"

a notion adopted from an assertion made by Vitruvius, as quoted by Matthioli (l. iii., c. 134), that in the island Crete, near the river Poterius, which flows between Gnosus and Cortyna, on the side towards Cortyna, the flocks and herds were found without spleens because they browsed on this herb; while, on the other side, towards Gnosus, they had spleens because it does not grow there. W. Coles, to improve the story, tells us that "if the asse be oppressed with melancholy he eates of this herbe, Asplenion or Miltwaste, and so eases himself of the swelling of the spleen." The notion was probably suggested, on the doctrine of signatures, by the lobular milt-like outline of the leaf in the species to which the name was originally given, the ceterach; a species which is now rather inconsistently made the type of a genus bearing this last name of "Ceterach," while another set of plants, in no respect resembling a spleen, are called "Spleenworts," and "Miltwastes." The enlarged spleen, called ague-cake, was that which it was supposed to waste or diminish when given medicinally. Gerarde and other herbalists praise its efficacy in all infirmities of this organ. W. Bulleyn indeed says (fol. 1) that " no herbe maie be compared therewith for his singular vertue to help the sickness or grief of the splene."

Asplenium, L.

MINT, L. *mentha*, Gr. μινθη.

„	BERGAMOT-,	M. citrata, Ehr.
„	BROOK-, HORSE-, or WATER-,	M. sylvestris, L.
„	PEPPER-,	M. piperita, L.
„	SPEAR-, or GARDEN-,	M. viridis, L.

MISTLETOE, A.S. *mistiltan*, from *mistl*, different, and *tan*, twig, being so unlike the tree it grows upon, a feature which Bacon in his Natural History (cent. vi. 556) has noted as its distinguishing character. " It is a plant," says he, " utterly differing from the plant upon which it groweth." By some it is derived from *mist*, dung, and much has been written in N. and Q. in support of this fanciful derivation. Viscum album, L.

MITHRIDATE MUSTARD, from being used in a medicine, named after Mithridates, a king of Pontus, who invented, as an antidote to all poisons, the famous preparation called after him *Mithridaticum*, into which this plant, among many more, was subsequently introduced. The original prescription, discovered by Pompey among the archives of the king, was very simple. Q. Serenus tells us that

> " Magnus scrinia regis
> Cum raperet victor, vilem deprehendit in illis
> Synthesin, et vulgata satis medicamina risit:
> Bis denum rutæ folium, salis et breve granum,
> Juglandesque duas, terno cum corpore ficus."

Other ingredients, animal as well as vegetable, were added to it from time to time, and the name changed to Theriaca. See TREACLE MUSTARD. Thlaspi arvense, L.

MITHRIDATE PEPPERWORT, Lepidium campestre, Br.

MOCK-PLANE, the sycamore, a translation of its Latin specific name, Acer Pseudoplatanus, L.

MOLY, the name of a plant in Homer's Odyssey, and occasionally introduced into modern poetry, as in Milton's Comus, l. 636, but not identified with any known species, and probably meant by Homer to be understood allegorically. See note on this subject in Hawkins's Milton, v. iv. p. 89.

MONEY-FLOWER, from its glittering round dissepiments left after the falling of the valves,   Lunaria biennis, L.

MONEY-WORT, from its round leaves,

Lysimachia Nummularia, L.

  ,,   CORNISH-, from its round leaves, and its growing in Cornwall,   Sibthorpia europæa, L.

MONKSHOOD, from the resemblance of the upper sepal to the cowl of a monk,   Aconitum Napellus, L.

MONK'S RHUBARB, a dock that, according to Tabernæmontanus (p. 824), was so called, " dieweil die wurzel der Rhabarbaren ähnlich ist, und von den Barfüssern und Carthaüsern in den klostern eine zeitlang heimlich gehalten ;" according to Parkinson, from its being the dock described as a rhubarb by the monks who commented upon Mesues.   Rumex Patientia, L.

MOON DAISY, a large daisy-like flower resembling the pictures of a full moon, the type of a class of plants, which, on the doctrine of signatures, were exhibited in uterine complaints, and dedicated in pagan times to the goddess of the moon and regulator of monthly periods, Artemis, whom Horsley (on Hosea ix. 10) would identify with Isis, the goddess of the Egyptians, with Juno Lucina, and with Eileithuia, a deity who had special charge over the functions of women ; an office in Roman Catholic mythology assigned to Mary Magdalene and Margaret. See MAUDLIN, MARGUERITE, and MAGHET.

Chrysanthemum leucanthemum, L.

MOON-WORT, a fern so called from the semilunar shape of the segments of its frond,   Botrychium Lunaria, Sw.

Moor-balls, from their globular form and occurrence in the lakes upon moors, Conferva ægagropila, L.

Moor-grass, Sesleria cærulea, Scop.

Moor-whin, or Moss-whin, a whin that grows on bleak heaths and mosses, Genista anglica, L.

Moor-wort, see Worts, Andromeda polifolia, L.

Morel, Fr. *morelle,* It. *morello,* dim. of *moro,* a Moor, L. *Maurus,* so called from its black berries.

,, Great-, the deadly nightshade,
Atropa Belladonna, L.

,, Petty-, the garden nightshade,
Solanum nigrum, L.

also a fungus, Fr. *morille,* Morchella esculenta, P.

Morgeline, from the French, L. *morsus gallinæ,* the henbit, Veronica hederifolia, L.

Moschatell, It. *moscatellina,* from L. *moschus,* musk, through *mosco, moscado,* musky, and its dim. *moscadello,* a plant so called from its faint musky odour,
Adoxa Moschatellina, L.

Moss, Fr. *mousse,* L. *muscus.*

,, Bog-, Sphagnum.
,, Cup-, Cenomyce pyxidata, Ach.
,, Iceland-, Cetraria islandica, Ach.
,, Irish-, Chondrus crispus, Lyng.
,, Rock-, Roccella tinctoria, Ag.
,, Water-, Fontinalis antipyretica, L.

Moss-berry, or Moor-berry, the cranberry, from its growing on moors or mosses, Vaccinium Oxycoccus, L.

Moss Campion, from its moss-like growth,
Silene acaulis, L.

Moss-crops, from *crop,* a head of flowers, and its place of growth. Ray and Plukenet (Alm. p. 201) say that in Westmoreland it is called by this name, because sheep are fond of it. It means merely moor-flowers.
Eriophorum vaginatum, L.

MOSS-RUSH, from its growing on heaths and mosses,
Juncus squarrosus, L.

MOTHER OF THOUSANDS (Treas. of Botany) a pun on its old name *penny-wort*,   Linaria cymbalaria, L.

MOTHER OF THYME, a name that undoubtedly ought to be written *Mother-Thyme*, as meaning "womb-thyme," having been given to this plant from its supposed effect upon the womb, which in our old writers is called *mother*, a use of the word adopted from the Flemish. According to Isidore (c. ix.) this plant was in Latin also called "*matris animula*, quod menstrua movet." Hence, too, its German names *Quendel* and *Kuttelkraut*. See HÖFER, v. ii., p. 184. Platearius says of it :

"Serpyllum matricem comfortat et mundificat. Mulieres Salernitanæ hoc fomento multum utuntur."

The nearly allied genus Melissa is, for the same reason, called by Herbarius (c. 84) *Muderkrut*, a word exactly equivalent to *Motherwort*. See below.
Thymus Serpyllum, L.

MOTHERWORT, so called, says Parkinson (Th. Bot. p. 44), from its being "of wonderful helpe to women in the risings of the *mother*,"   Leonurus Cardiaca, L.

„   also in old works the mugwort, which, from its being used in uterine diseases, was called *moder-wort*, womb-wort, a name that by Ælfric is correctly rendered *matrum-herba*, wort of mothers, but by later writers misunderstood and rendered *mater herbarum*, mother of worts. Thus Macer (c. i.)—

"*Herbarum matrem* justum puto ponere primo :
Præcipue morbis muliebribus illa medetur."

Indeed its use in these affections was so general that Ray tells us (Cat. Plant. p. 29), quoting the words of Schröder :—" Uterina est, adeoque usus est creberrimi mulierculis, quæ eam adhibent interne et externe, ut vix

balnea et lotiones parent in quibus artemisia non conti-
neatur." Like other uterine herbs it was dedicated to
the goddess Artemis, and thence its Latin name.

Artemisia vulgaris, L.

MOULD, in ink and other fluids, usually

Hygrocrocis, Ag.

MOULDINESS, Aspergillus, Mich.

MOUNTAIN ASH, from its pinnate leaves called an *Ash*,
the wild service tree, Pyrus Aucuparia, L.

MOUNTAIN COWSLIP, Primula Auricula, L.

MOUNTAIN ELM, the wych elm, Ulmus montana, L.

MOUNTAIN FERN, Aspidium Oreopteris, Sw.

MOUNTAIN SORREL, Oxyria reniformis, L.

MOUSE-BARLEY, G. *maus-gerste*, Sw. *mus-korn*, a translation of its Latin name, which was given to it either from
the belief that it was the *lolium murinum* of Pliny (l. xxii.
c. 25), or, through a confusion between *murinum*, of a
mouse, and *murale*, of a wall, to express that it was a *wall-*
barley; and this last is the most probable origin of the
name, since Tragus tells us that it was called so, because
it grows upon *walls*: " weil es von sich selbst auf den
Mauren wächst;" and Dr. Wm. Turner (pt. ii. p. 17):
" that it was called of the Latines *Hordeum murinum*,
that is wall-barley." Hordeum murinum, L.

MOUSE-EAR, from the shape of the leaf,

Hieracium Pilosella, L.

MOUSE-EAR CHICKWEED, Cerastium vulgare, L.

MOUSE-EAR SCORPION-GRASS, the plant now called " Forget-me-not," from its one-sided raceme being curved like
that creature's tail, and its small soft oval leaves,

Myosotis palustris, L.

MOUSE-TAIL, from its slender cylindrical seed-spike,

Myosurus minimus, L.

MOUSE-TAIL GRASS, Martyn in Fl. rustica, from the
shape of the spike, Alopecurus agrestis, L.

MOUTAN, from the Chinese *Meu-tang*, king of flowers, the tree peony,   Pæonia Moutan, L.

MUDWORT, from its place of growth,
            Limosella aquatica, L.

MUGGET, Fr. *muguet*, O.Fr. *muzquet*, from L. *muscatus*, scented with musk, a name applied in French to several flowers, and to the nutmeg as *noix muguette*, in English to the lily of the valley, Fr. *muguet de Mai*,
            Convallaria majalis, L.

MUGGET, PETTY-, Fr. *petit-muguet*, little dandy, a word applied to effeminate dressy young men, Jemmy Jessamies, with puffy yellow hair,   Galium verum, L.

MUG-WEET, GOLDEN-, a corruption of Fr. *muguet*,
            Galium cruciatum, DC.

MUGWORT, a name that corresponds in meaning with its synonym *wyrmwyrt*, wormwood, from O.E. *mough*, *moghe*, or *moughte*, a maggot or moth, a word used by Hampole (P. o. C. 1. 5572) :

> "And wormes and *moghes* on þe same mancre
> Sal þat day be in wittenes broght;"

and by Wycliffe (Matt. vi. 20) :

> "Where neþer ruste ne *moughte* destruyeþ ;"

a name given to this plant from its having been recommended by Dioscorides to ward off the attacks of these insects, whence Macer (c. 3) de Absinthio :

> "A tineis tutam reddit qua conditur arcam."

and Wm. Bulleyn, speaking of wormwood, says, fol. 2 :

> "It kepeth clothes from wormes and mothes."

The name is explained by an old writer in MS. Arundel, 42, fol. 35, as a form of *Motherwort*. "Mogwort, al on as seyn some, modirwort : lewed folk þat in manye wordes conne no rygt sownynge, but ofte shortyn wordys, and changyn lettrys and silablys, þey corruptyn þe o in to u, and d into g, and syncopyn i, smytyn awey i and r, and seyn mugwort." It is unnecessary to have recourse to this singular process. The plant was known both as a *moth-wort* and as a *mother-wort*, but while it was used almost exclusively as a *mother-wort*, it still retained, at the same time, the name of *mug-wort*, a synonym of *moth-wort*. In Ælfric's glossary it is called *matrum herba.*

Artemisia vulgaris, L.

MULBERRY, by a change of *r* to *l*, from L. *morus*, Gr. μορον, a word of unknown origin, which was introduced into Greece with the tree, M. nigra, L.

MULLEIN, or WHITE MULLEIN, in old works *Molayne*, A.S. *molegn*, the hig-taper, Fr. *moleine*, the scab in cattle, O.Fr. *malen*, L. *malandrium*, the malanders or leprosy, whence *malandrin*, a brigand, from lepers having been driven from society, and forced to a lawless life. The term *malandre* was applied to other diseases of cattle, to lung diseases among the rest, and Marcellus Empiricus explains it as "morbus jumenti quo tussit." The hig-taper, being used for these, acquired its names of *Mullein*, and bullock's lungwort. Verbascum Thapsus, L.

„ PETTY-, the cowslip. "Those herbes," says Gerarde, "which at this day are called Primroses, Cowslips, and Oxelips, are reckoned among the kinds of Mulleins, for that the ancients have named them *Verbasculi*, that is to saie, small Mulleins." Primula veris, L.

MULLET, FLEABANE-, a plant used to destroy fleas, and called *mullet*, Fr. *mollet*, from its soft leaves,

Inula dysenterica, L.

MUSCOVY, or MUSK, from its odour,

Erodium moschatum, L'Her.

MUSHROOM, Fr. *mouscheron*, at present spelt *mousseron*, a name applied to several species of Agaricus, and derived by Diez from *mousse*, moss, with which it is difficult to see how mushrooms are connected. One of the most conspicuous of the genus, the A. muscarius, is used for the

11

destruction of flies, *mousches*, and as Albertus Magnus says (l. vii. 345):

"Vocatur *fungus muscarum*, eo quod in lacte pulverizatus interficit muscas."

and this seems to be the real source of the word, which, by a singular caprice of language, has been transferred from this poisonous species to mean, in the popular acceptation of it, the wholesome kinds exclusively.   Agaricus, L.

MUSK ORCHIS, from its scent,

Herminium monorchis, RB.

MUSK THISTLE, from its scent,   Carduus nutans, L.

MUSTARD, according to Diez, from L. *mustum*, new wine, which he says is used in preparing it.   It seems far more likely to be the Sp. *mastuerzo*, from L. *nasturtium*, cress, so called, it is said, from *nasitortium*, *a naso torquendo*, alluding to the wry faces and sneezing that it causes.

		Brassica alba, Bois.
,,	BLACK-,	Brassica nigra, Bois.
,,	BOWYERS-,	Lepidium ruderale, L.
,,	GARLIC-,	Erysimum Alliaria, DC.
,,	HEDGE-,	Sisymbrium officinale, L.
,,	MITHRIDATE-,	Thlaspi arvense, L.
,,	TOWER-,	Turritis glabra, L.
,,	WILD-, see CHARLOCK,	Brassica Sinapistrum, Bois.

MYPE, Wel. *maip*, Gael. *neip*, given in Gerarde (p. 871) as a name of the parsnip, a corruption of L. *napus*, and properly meaning the turnip,   Brassica Rapa, L.

MYRTLE, It. *mirtillo*, dim. of *mirto*, L. *myrtus*, Gr. μυρτος,   Myrtus communis, L.

NAILWORT, perhaps more correctly *Agnail-wort*, the whitlow-grass, from its supposed curative powers, in cases of agnail,   Draba verna, L. and Saxifraga tridactylites, L.

NAKED LADIES, G. *nakte jungfer*, from the pink flowers rising naked from the earth, the meadow saffron,

Colchicum autumnale, L.

NANCY PRETTY, or NONE-SO-PRETTY, an unexplained name of the London pride, Saxifraga umbrosa, L.

NAP-AT-NOON, from it flowers closing at midday, the goat's beard, Tragopogon porrifolius, L.

NARCISSUS, Gr. ναρκισσος, from ναρκαω, become numb, related to Skr. *nark*, hell, so called from the torpidity caused by the odour of the flower, as remarked by Plutarch, who (in Sympos. con. 3, c. 1) says : τον ναρκισσον, ὡς ἀμβλυνοντα τα νευρα και βαρυτητας ἐμποιουντα ναρκωδεις· διο και ὁ Σοφοκλης αὐτον ἀρχαιον μεγαλων θεων στεφανωμα (τουτεστι των χθονιων) προσηγορευκε : "Narcissus, as blunting the nerves, and causing narcotic heaviness : wherefore also Sophocles called it the ancient chaplet of the Great (that is the Infernal) gods." The passage is quoted from an exquisite chorus of the Œdipus at Colonos, where (at l. 682) the original has μεγαλαιν θεαιν, the two great goddesses, meaning Ceres and Proserpine. The epithet which the poet here applies to the narcissus, καλλι-βοτρυς, finely clustered, suggests that he meant the hyacinth, a plant which, from its heavy odour and dark colour, was more likely than the one we now call narcissus to have been consecrated to those deities. Plutarch adds that, " those who are numbed with death should very fittingly be crowned with a benumbing flower." The coincidence of the name *narcissus* with the Skr. *nark* indicates some very ancient traditionary connexion of Greek with Asiatic mythology. Ovid, who undoubtedly means one of the plants which still bear this name, represents it as having been so called after a youth who pined away for love of his own image reflected in a pool of water ; an instance, among many more, of a legend written to a name ; for as an old poet, Pamphilus, remarks, Proserpine was gathering Narcissi long before that youth was born. Narcissus, L.

NARD, Gr. ναρδος, the name of various aromatic plants, chiefly of the valerian tribe, that were formerly, and are still used in Asiatic harems.

NAVEL-WORT, from the shape of its leaf,
Umbilicus pendulinus, DC.

NAVEW, Fr. *naveau*, from *napellus*, dim. of *napus*, the
rape, Brassica Napus, L.

NECKWEED, a cant term for hemp, as furnishing halters
for the necks of criminals, Cannabis sativa, L.

NECTARINE, It. *nettarino*, dim. of *nettare*; L. *nectar*, Gr.
νεκταρ, the drink of the gods, and called so from its flavour,
Amygdalus persica, var. lævis, L.

NEELE, found in old books as a translation of Gr. ζιζανια,
and equivalent to cockle or darnel, Fr. *nielle*, L. *nigella*,
blackish, once used to mean weeds generally, but in later
works restricted to the larger ray grass.

"Frumentis nocuam lolium Græcus vocat herbam,
Quam nostri dicunt vulgari more *nigellam*." Macer, c. 64.
Lolium temulentum, L.

NEEDLE FURZE, from its delicate spines,
Genista anglica, L.

NEP or NEPPE, contracted from L. *nepeta*,
Nepeta cataria, L.

NETTLE, A.S. and Du: *netel*, Da. *naelde*, Sw. *naetla*,
G. *nessel*, the instrumental form of *net*, the passive participle of *ne*, a verb common to most of the Ind-European
languages in the sense of "spin" and "sew," Gr. νεειν,
L. *ne-re*, G. *nä-hen*, Skr. *nah*, bind. *Nettle* would seem to
have meant primarily that with which one sews. Applied to
the plant now called so, it indicates that this supplied the
thread used in former times by the Germanic and Scandinavian nations, which we know as a fact to have been the
case in Scotland in the seventeenth century. Westmacott
says (p. 76) "Scotch cloth is only the housewifery of the
nettle." In Friesland also it has been used till a late
period. Flax and hemp bear southern names, and were
introduced into the North to replace it. Urtica, L.

„ BEE-, Galeopsis versicolor, Curt.

Nettle, Dead-, Lamium, L.

   ,, Hedge-, from its nettle-like leaves and place of growth, more properly *Hedge Dead-nettle*,

                  Stachys sylvatica, L.

   ,, Hemp-, Galeopsis, L.

   ,, Roman-, from being found abundantly about Romney in Kent, and the report that " Roman soldiers brought the seed with them, and sowed it there for their own use, to rub and chafe their limbs, when through extreme cold they should be stiffe and benummed ; having been told that the climate of Britain was so cold, that it was not to be endured without some friction or rubbing, to warm their bloods and to stir up natural heat." Park-kinson (Th. Bot. p. 441). Lyte's explanation of this and other applications of the term " Roman" is more probable. " It is a straunge herbe, and not common in the countrey, and they do call al such straunge herbes as be unknowen of the common people, Romish or Romayne herbes, although the same be brought from Norweigh."

                  Urtica pilulifera, L.

Nightshade, A.S. *niht-scada*, O.H.G. *naht-scato*, from its officinal Lat. name *solatrum*, which is derived, as an instrumental noun, from L. *solari*, soothe, as *aratrum* from *arare*, and means " anodyne." Under this form we find it in the Ort. San. (c. 349) ; and Matthioli, in speaking of the Belladonna (c. 59), describes it as " eam plantam quam herbariorum vulgus *solatrum majus nominat*." This word *solatrum* has been mistaken for *solem atrum*, a black sun, an eclipse, a shade as of night. Solanum, L.

   ,, Bittersweet-, see Bitter-sweet,

                  Solanum Dulcamara, L.

   ,, Deadly-, Atropa Belladonna, L.

   ,, Enchanter's-, Circæa Lutetiana, L.

   ,, Woody-, the bittersweet.

Ninety-knot, see Knot-grass and Centinode.

NIPPLE-WORT, Fr. *herbe aux mamelles*, from its use in cases of sore nipple, Lapsana communis, L.

NIT-GRASS, from its little nit-like flowers, a translation of its L. specific name, *lendigerum*,

Gastridium lendigerum, L.

NONE-SO-PRETTY, or NANCY-PRETTY, the London pride, or Pratling parnel, terms that seem to allude to the heroine of some popular farce, song, or tale, Saxifraga umbrosa, L.

NONSUCH, " a name conferred upon it from its supposed superiority as fodder." Smith in Eng. Bot.

Medicago lupulina, L.

in Gerarde and Parkinson applied to the scarlet lychnis,

Lychnis chalcedonica, L.

NOON-FLOWER, or NOON-TIDE, from its closing at mid-day, and marking the hour of noon,

Tragopogon pratensis, L.

NOOPS, i.e. *knops*, A.S. *cnæp*, a button, a name of the cloudberry used on the Eastern Border,

Rubus Chamæmorus, L.

NOSEBLEED, the yarrow, from its having been put into the nose, as we learn from Gerarde, to cause bleeding and to cure the megrim, and also from its being used as a means of testing a lover's fidelity. Forby in his East Anglia (p. 424) tells us that in that part of England a girl will tickle the inside of the nostril with a leaf of this plant, saying,

"Yarroway, yarroway, bear a white blow;
If my love love me, my nose will bleed now."

Parkinson (Th. Bot. p. 695) says that "it is called of some *Nose-bleede* from making the nose bleede, if it be put into it, but assuredly it will stay the bleeding of it." This application of the yarrow, and all the superstitions connected with it, have arisen, as in so many other instances, from the medieval herbalists having been misled by a name, and taken one plant for another. Isidore (c. ix.) in speaking of a polygonum, but meaning by that name a horse-tail, says that it was called *herba sanguinaria*, from its being used to make the nose bleed. Apuleius calls the horsetail *millefolium*, and this term *millefolium* was subsequently transferred to the yarrow, which acquired the names of *Herba sanguinaria* and *Nose bleed*, and with the names the remedial character of the horse-tail, and its superstitious appliances. See SANGUINARY.

Achillæa Millefolium, L.

NOSTOC, some alien word, the name of a genus of Algæ so called. See FALLEN STARS. Tremella Nostoc, L.

NUT, A.S. *hnut*, Ic. *hnitt*, Sw. *nott*, Da. *nödd*, G. *nuss*, L. *nux*, words connected with *knit*, *knot*, *knopf*, *knob*, implying a hard round lump.

,,	BLADDER-, from its inflated capsule,	
		Staphylea pinnata, L.
,,	CHEST-, .	Castanea vesca, DC.
,,	EARTH-, or PIG-, or JUR-, or HOG-,	
		Bunium flexuosum, With.
,,	FRENCH-, the walnut,	Juglans regia, L.
,,	HAZEL-, or WOOD-,	Corylus Avellana, L.
,,	WAL-,	Juglans regia, L.

OAK, A.S. *ac*, *æc*, Scot *aik*, O.N. *eik*, Sw. *ek*, Da. *eg*, Ic. *eyk*, L.G. *eek* and *eik*, G. *eiche*, O.H.G. *eih*, the *h* having a guttural sound. All these words refer to the fruit of the tree, the acorn, from which, as its most useful product, the oak took its name. "During the Anglo-Saxon rule," says Selby, p. 227, "and even for some time after the Conquest, oak forests were chiefly valued for the fattening of swine. Laws relating to pannage, or the fattening of hogs in the forest, were enacted during the Heptarchy, and by Ina's statutes any person wantonly injuring or destroying an oak-tree was mulcted in a fine varying according to its size, or the quantity of mast it produced." Quercus, L.

OAK OF CAPPADOCIA, or -OF JERUSALEM, from a fancied resemblance of its leaf to that of an oak, and its coming from a foreign country,   Chenopodium ambrosioides, L.

OAK-FERN, of old herbals,   Polypodium vulgare, L.

„ of modern botanists,

               Polypodium Dryopteris, L.

OAT, A.S. *ata*, a word that seems originally to have meant "food," the O.N. *áta*, and Lat. *esca*, for *edca* or *etca*, and derived from words signifying "eat," A.S. *etan*, Lat. *edere*, from an ancient root, the Skr. *ad*, and applied to the oat exclusively, as being once the chief food of the north of Europe. With this word *ata* is etymologically connected, and indeed, identical, G. *aas*, a carcase, the term having, apparently, been adopted, in the former sense by an agricultural, and in the latter by a carnivorous, a shepherd or hunter tribe of the Germanic race : an evidence, as far as it goes, that we must not assume our various dialects to have originated simultaneously from any one common tongue, or in any one district.   Avena sativa, L.

„ WILD-,   Avena fatua, L.

OAT-GRASS, a farmer's term, according to Martyn in Fl. Rust., but certainly not a common one, for

              Bromus mollis, L.

OFBIT, in Turner OFBITEN, for *bitten-off*, the Devil's bit, from the appearance of the root,   Scabiosa succisa, L.

OIL-SEED, from oil being made from it,

              Camelina sativa, L.

OLD-MAN, southernwood, from its use as recommended by Pliny (l. xxi. c. 21), and as explained in the line of Macer, c. ii.:

    "Hæc etiam venerem, pulvino subdita tantum,
    Incitat."

             Artemisia Abrotanum, L.

OLD-MAN'S-BEARD, from its long white feathery awns, the traveller's joy,   Clematis Vitalba, L.

ONE-BERRY, from its one central fruit, the trulove,
Paris quadrifolia, L.

ONE-BLADE, from its barren stalk having only one leaf.
Its Latin specific name implying "two-leaved" refers to
the flowering stalk. Maianthemum bifolium, DC.

ONION, Fr. *oignon*, in a Wycliffite version of Num. xi. 5,
uniowns, from L. *unio*, some species of it mentioned by
Columella, Allium Cepa, L.

,, WELSH-, not from Wales, but the G. *wälsch*,
foreign, the plant having been introduced through Germany from Siberia, Allium fistulosum, L.

ORACH, formerly *Arach*, in Pr. Pm. *Arage*, in MS. Harl.
978, *Arasches*, Fr. *arroche*, a word that Menage and Diez
derive from L. *atriplice*. Its Gr. name χρυσολαχανον,
golden herb, suggests a more probable explanation of it in
a presumed M.Lat. *aurago*, formed from *aurum*, gold, by
the addition to it of *ago*, wort, as in plantago, lappago,
solidago, etc., and this word *aurago* would become in French
arroche, as borago *bourroche*. At the same time its use in
the cure of jaundice, *aurugo*, may have fixed upon the
plant the name of the disease.

> "*Atriplicem* tritam cum nitro, melle, et aceto,
> Dicunt appositam calidam sedare podagram:
> *Ictericis* dicitque Galenus tollere morbum
> Illius semen cum vino sæpius haustum."
Macer, c. xxviii. l. 7.
Atriplex hortensis, L.

ORCHANET, from the French. See ALCANET.

ORCHARD-GRASS, from its growing in orchards under the
drip of trees, Dactylis glomerata, L.

ORCHAL, ORCHEL, or ORCHIL, the rock-moss, supposed
by Scheler to be a transposition of *rochelle*, a small rock,
Roccella tinctoria, Ag.

ORCHIS, Gr. ὀρχις, from its double tubers,

,, BEE-, from the resemblance of its flowers to a bee,
Ophrys apifera, L.

ORCHIS, BOG-,	Malaxis paludosa, Sw.
„ BUTTERFLY-,	Habenaria bifolia, RB.
„ DRONE-,	Ophrys fucifera, Sm.
„ FLY-,	Ophrys muscifera, Huds.
„ FROG-,	Habenaria viridis, RB.
„ GREEN-MAN-,	Aceras anthropophora, RB.
„ GREEN MUSK-,	Herminium monorchis, RB.
„ HAND-,	Orchis maculata, L.
„ LIZARD-,	Orchis hircina, Scop.
„ MAN-,	Aceras anthropophora, RB.
„ MILITARY-,	Orchis militaris, L.
„ MONKEY-,	Orchis tephrosanthos, Vill.
„ MUSK-,	Herminium monorchis, RB.
„ SPIDER-,	Ophrys aranifera, Hud.
	and arachnites, Willd.

ORGANY, or ORGAN, marjoram, from L. *origanum*, Gr. ὀρίγανον, Origanum vulgare, L.
also the penny-royal, Mentha Pulegium, L.

ORPINE, Fr. *orpin*, contracted from *orpiment*, L. *auripigmentum*, gold pigment, a sulphuret of arsenic, a name given in old works to certain yellow-flowered species of the genus, but, perversely enough, transferred of late to almost the only European one that has pink flowers,

Sedum Telephium, L.

ORRICE, either from its official Latin name, *Acorus Dioscoridis*, or from *Ireos*, (sc. radix) by transposition of the vowels, or very probably from some confusion between these two words, *ireos* and *acorus*; since the roots of two different species of Iris were known as *Acorus falsus* or *adulterinus*, and sold for those of Acorus Calamus, L. (Bauhin's Pinax. p. 34.) It cannot be derived, as in our dictionaries, from *Iris*, the initial *I* of which could not have become *O*, and could scarcely have remained unaspirated. At present it means the Florentine Iris, but is used in older works as a generic name, and in Cotgrave, and old

German herbals, applied as *Wild Ireos*, to the water flower de luce, and to the stinking gladdon. Iris, L.

OSIER, Fr. *osier*, M.Lat. *oseria*, whence *oseretum*, a withy-bed, from a Celtic word meaning water, or *ooze*, that has given its name to the Oise in France, and to several rivers in England, spelt according to the dialect of the district, Ouse, Ose, Use, or Ise, and which in M.Lat. would have made *Osa*, whence an adjective *osaria*, aqueous, and *osier*. Salix viminalis, L.

OSMUND, OSMUND ROYAL, or OSMUND THE WATERMAN, apparently a corruption of G. *gross mond-kraut*, greater moon-wort, representing its ancient officinal name *lunaria major*. There are other derivations of it, such as that by Beckmann, from the name of some person ; by Nemnich, on the authority of Houttuyn, from *os*, mouth, and *mundare*, cleanse; by others from *os*, bone, and *mundare*, cleanse. The *Waterman* would seem to be its Flemish name, *Water-varn*. The *Royal* refers, we are told by Lobel (Kruydb. i. p. 991), to its great and excellent virtues.

Osmunda regalis, L.

OSTERICK, M.Lat. *ostriacum*, apparently a corruption of L. *aristolochia*, a name transferred to it from another plant, Polygonum Bistorta, L.

OUR LADY'S BEDSTRAW, etc. See LADY'S.

OWLER, a corruption of *Aller*, the alder tree.

OX-EYE, the great daisy, a translation of L. *buphthalmus*, Gr. βουφθαλμον, a name now appropriated to a different genus, Chrysanthemum Leucanthemum, L.

OX-HEEL, or more properly OX-HEAL, A.S., *oxnalib*, from its being used in settering oxen. See SETTERWORT.

Helleborus fœtidus, L.

OXLIP, A.S., *oxan-slippe*, a word that, like *Cowslip*, is of very uncertain derivation. O. Cockayne (in Leech. ii. p. 378) suggests that the second syllable may be *slyppa*, a soft viscid mass, but leaves unexplained what this has to do with the plant. Primula veris caulescens, L. elatior, Jacq.

OXTONGUE, from the shape and roughness of its leaf,
Helminthia echioides, Gärt.

OYSTER-GREEN, a sea-weed, so named from its bright green tint, and its being frequently found attached to the oyster, Ulva lactuca, L.

PADDOCK-PIPES, in Cotgrave TOAD-PIPES, from its straight

hollow pipe-like stalks, and growth in mud, where toads haunt, the horse-tail,   Equisetum limosum, L.

PADDOCK-STOOLS, in Topsell PADSTOOLE, Du. *padde-stoel*, toad-stool, from their resemblance to the tripods called joint-stools, and the notion that toads sit upon them. (See TOADSTOOL).   Boletus and Agaricus.

PADELION, Fr. *pas de lion*, from the resemblance of its leaf to the impress of a lion's foot, the lady's mantle,
Alchemilla vulgaris, L.

PAIGLE, PAGLE, PAGEL, PEAGLE, PEGYLL, and PYGIL, a name that is now scarcely heard except in the Eastern counties, and usually assigned to the cowslip, but by Ray and Moore to the Ranunculus bulbosus, a word of extremely obscure and disputed origin. Most of the dictionaries derive it from *paralysis*. Latham from Fr. *épingle*, a pin, in allusion to its pin-shaped pistil; Forby, strangely enough, from A.S. *paell*, a die-plant, a purple robe; Forster, in Perennial Calendar, p. 191, says that it " evidently signifies *pratingale*, from *prata*, meadows, where it delighteth to grow." An East Anglian correspondent informs me that *paigle* means a spangle. In Flemish *pegel* is a gauge. It is possible that it may be corrupted from A.S. *cæg*, a key, or from some word compounded with it. The primroses and mulleins are so mixed up together by the herbalists, that I rather incline to the belief that it is a name of the mullein under which cowslips and primroses were comprehended, and that it is not descriptive of these latter. It may possibly be a corruption of *verbasculum*, through a lost French word, in which the *s* was omitted before *c*.   Primula veris, L.

PALM, L. *palma*, Gr. παλαμη, the palm of the hand, from the shape of the leaf in the species most familiar to Greek and Latin writers, the dwarf palm of the south of Europe, a name given in England to the sallow with its catkins in flower, from its branches having formerly been carried in processions, and strown on the road the Sunday next before Easter, in imitation of the palm leaves that were strown before Jesus on his entry into Jerusalem. A representative of the Eastern tree was required, and these golden tassels presented themselves at precisely the right season. Branches of the yew tree, on account of its leaves being green at this season, were also used. In an old sermon on Palm Sunday, quoted by Hampson, (ii. 300) the account of it is as follows : "þan Ihu yode towerde Jerusalem, and þe pepul brokon brawnches of olyfe and of palme and keston in þe way, &c., but for encheson we have non olyfe þat bereth grene leves, we takon in stede of hit *hew* and *palmes wyth*, and bereth abowte in procession." (Cott. MS. Claud. A. 11. fo. 52.) This was the ancient usage in Scotland, as described by Sir Walter Scott in his Castle Dangerous : " Several of the Scottish people, bearing willow branches, or those of yew, to represent the palms which were the symbol of the day [Palm Sunday] were wandering in the churchyard." It was the custom also in East Kent, according to Evelyn's Sylva ; in Dorsetshire (Notes and Queries, 3 S. vii. p. 364) ; and in Ireland ; and the yew-tree which, as well as the willow, was popularly called "Palm," was planted in churchyards to supply boughs for these occasions.

Salix caprea, L., and Taxus baccata, L.

PALSY-WORT, L. *Herba paralyseos*, from its supposed power to cure the palsy, the cowslip, Primula veris, L.

PANCE or PAUNCE, see PANSY.

PANICK-GRASS, L. *panicum*, which Pliny says was "a paniculis dictum," so called from its panicles. The word

seems to be formed from *panus*, a head of millet, and to be
connected with *panis*, bread, from an ancient root *pa*, feed,
retained in *pa-sco*, *pa-bulum*, and *pa-ter*. See Bopp. comp.
Gram. p. 1164.   Panicum, L.

PANSY, or PAUNCE, Fr. *pensée*, thought, once called *menues
pensées*, It. *pensieri menuti*, idle thoughts, G. *unnütze sorge*.
Dr. Johnson and Talbot would derive the name from
L. *panacea*, but the plant has never been called so, nor
regarded as a panacea.   Its habit of coquettishly hanging
its head, and half hiding its face, as well as some fancied
resemblances in the throat of the corolla, has led to many
quaint names in our own, and in foreign languages : " Cull
me-," or " Cuddle me to you," " Love and idle," " Live in
idleness," or " Love in idleness," a line, perhaps, of some
song or poem, " To live and love in idleness," but originally, it would seem, " Love in idle," that is, " in vain,"
and in Lobel, " Love in idle Pances," " Tittle my fancy,"
" Kiss me, ere I rise," " Jump up and kiss me," " Kiss me
at the garden gate," " Pink of my John," and several
more of the same amatory character.   From its three
colours combined in one flower, it is called " Herb Trinity,"
and " Three faces under a hood ;" from confusion with the
wallflower, " Heartsease;" and from M. Lat. viola flammea,
" Flame flower."   There is no plant, except, possibly, the
ground ivy, that has obtained so many names, and curious
sobriquets.   Viola tricolor, L.

PARIS, see HERB PARIS.

PARK-LEAVES, a name that seems, like its Danish,
Swedish, and Norwegian synonym, *pirkum* or *perkum*, to
have been suggested by L. *Hypericum*, Gr. ὑπερικον, but
taken in the sense of *perked* or pricked leaves, from those
of the commonest species of the genus, H. perforatum, L.
being so dotted with resinous deposits, as to look as if they
were pricked all over; a character not observable in that
to which the name of *Park-leaf* is now restricted.   Its

French synonym, *parcoeur*, by heart, seems, like the English name, to have been suggested by the Latin, through an accidental coincidence of sound. H. Androsæmum, L.

PARNASSUS GRASS, a plant supposed to be one described by Dioscorides as growing on Mount Parnassus,

Parnassia palustris, L.

PARSLEY, spelt in the Grete Herball *Percely*, Fr. *persil*, L. *petroselinum*, from Gr. πετρος, rock, and σελινον, some umbelliferous plant,   P. sativum, Koch.

,,	BASTARD-, or BUR,	Caucalis daucoides, L.
,,	COW-,	Chærophyllum sylvestre, L.
,,	FOOL'S-,	Œthusa Cynapium, L.
,,	HEDGE-,	Caucalis Anthriscus, Huds.
,,	MILK-,	Peucedanum palustre, Mn.
,,	STONE-,	Sison Amomum, L.

PARSLEY-FERN, from the resemblance of its fronds to parsley leaves,   Cryptogramma crispa, R.B.

PARSLEY-PIERT, or PARSLEY-BREAK-STONE, Fr. *percepierre*, of *percer*, pierce, and *pierre*, stone, from its being used in cases of stone in the bladder, and so called, according to W. Coles (Ad. in Ed. ch. 222), "from its eminent faculties to that purpose,"   Alchemilla arvensis, Sm.

PARSNIP, or, as it is spelt in old herbals, PASNEP and PASTNIP, from L. *pastinaca*, by change of *c* to *p*,

Pastinaca sativa, L.

,,	COW-,	- Heracleum Sphondylium, L.
,,	WATER-,	Sium latifolium, L.

PASQUE- or PASSE-FLOWER, Fr. *pasques*, Gr. πασχα, Heb. *pesach*, a crossing over, from its blossoming at Easter, that in old works was called *Pask*, as in Robert of Brunne, p. 263 :

"Fro gole to þe *pask*, werred Sir Edward,"

and *Passe* or *Pase*, as in Levin's Manip. col. 36.

Anemone Pulsatilla, L.

PASSIONS, or PATIENCE, a dock so called, apparently,

from the Italian name under which it was introduced from the South, *Lapazio*, a corruption of L. *lapathum*, having been mistaken for *la Passio*, the Passion of Jesus Christ,   Rumex Patientia, L.

PAUL'S BETONY, a name given to it by Turner, as being the plant described as a betony by Paul Ægineta,
                        Veronica serpyllifolia, L.

PAWNCE, in Spenser, the Pansy.

PEA, in old works PEASE, a word that has either arisen from Fr. *pois*, pronounced, as it used to be, *pay*, or from the old form *pease* being, like *cerise*, a cherry, mistaken for a plural.   The Lat. *pisum*, from which it is derived, means brayed in a mortar, *pinsum*, or, as it is spelt in Apuleius, *pisatum*, Gr. πισος, from Skr. *pish*, bray, whence *peschana*, a quern or handmill.   Tusser makes the plural *peason* agreeably to a practice of ending the plural with *n*, when the singular ends with *s*, as e.g. oxen, housen, hosen, from ox, house, hose.   Pisum sativum, L.

,,	CHICK-,	Cicer arietinum, L.
,,	CHICKLING-,	Lathyrus, L.
,,	EVERLASTING-,	Lathyrus latifolius, L.
,,	HEATH-,	Orobus tuberosus, L.
,,	SWEET-,	Lathyrus odoratus, L.
,,	WOOD-,	Orobus tuberosus, L.

PEACH, in old works spelt PESKE, PEESK, PESHE, and PECHE, O.Fr. *pesche*, L. *persica*, formerly called *malum persicum*, Persian apple, from which the Arabs formed their name for it with the prefix *el* or *al*, and thence the Spaniards *alberchigo*,   Amydalus Persica, W.

PEACH-WORT, from the resemblance of its leaves to those of the peach,   Polygonum Persicaria, L.

PEAR, a foreign word adopted from the Southern into the Germanic languages, It. and Sp. *pera*, Fr. *poire*, probably once pronounced *paire*, from L. *pyrus*,
                        Pyrus communis, L.

PEARL-PLANT, from its smooth hard pearly seed,
Lithospermum officinale, L.

PEARL-GRASS, from its glittering panicles,
Briza maxima, L.

PEARL-WORT, from its being used to cure a disease of the eye called pearl, Sagina, L.

PEASELING, an inferior *Pea*, (compare Chickling, Vetchling, Crambling), Orobus, L.

PEGROOTS, (Dale, p. 177) the green hellebore, from its roots being used by cattle doctors in the operation of pegging or settering. See SETTERWORT.
Helleborus viridis, L.

PELL-A-MOUNTAIN, or PENNY MOUNTAIN, corruptions of ser*pyllum montanum*, hill thyme, Thymus Serpyllum, L.

PELLITORY, or PARITORY, OF THE WALL, L. *parietaria*, from *paries*, a house-wall, into which this weed usually grows, Parietaria officinalis, L.

PELLITORY, or PELLETER OF SPAIN, Sp. *pelitre*, L. *pyrethrum*, Gr. πυρεθρον, " by reason of his hot and fiery taste," says Gerarde, p. 758. The term *Pellitory of Spain* seems merely to refer to its being the plant called so in Spain, and not to its being brought thence.
Anacyclus Pyrethrum, DC.

PENNY-CRESS, from its round flat silicules, resembling silver pennies, Thlaspi arvense, L.

PENNY-GRASS, from its round seeds like silver pennies,
Rhinanthus Crista galli, L.

PENNY-ROT, in Lyte PENNY-GRASS, from its character of giving sheep the rot, and its small round leaves,
Hydrocotyle vulgaris, L.

PENNY-ROYAL, from L. *puleium regium*, through Du. *poley*, in the old herbals called *puliol royal;* its Latin name being derived from its supposed efficacy in destroying fleas, *pulices*, Pliny (b. xx. c. 54).
Mentha Pulegium, L.

PENNY-WORT, from its round leaves,
>> Sibthorpia europæa, L.
,, in old works Linaria Cymbalaria, L.
,, MARSH-, Hydrocotyle vulgaris, L.
,, WALL-, Cotyledon Umbilicus, L.

PEONY, or PIONY, L. *Pæonia*, Gr. παιωνια, from Παιων, a god of physic, supposed to be the same as Apollo, who healed the gods Ares and Hades of their wounds (Hom. Il. v. 401 and 899), Pæonia corallina, Retz.

PEPPER, L. *piper*, Gr. πιπερι, Skr. *pippali*.
,, WALL-, from its biting taste, and place of growth,
>> Sedum acre, L.
,, WATER-, Polygonum Hydropiper, L.
and also Elatine Hydropiper, L.

PEPPER-CROP, a cyme or head of flowers with the pungent taste of pepper, the stone-crop, Sedum acre, L.

PEPPER-GRASS, a plant with linear grass-like leaves, and pepper-corn-like pellets of inflorescence,
>> Pilularia globulifera, L.
PEPPER-MINT, Mentha piperita, L.
PEPPER SAXIFRAGE, Silaus pratensis, L.

PEPPER-WORT, from their acrid taste, the cresses, but more particularly Lepidium latifolium, L.

PERCEPIER, Fr. *percepierre*, pierce-stone, from its supposed lithontriptic virtues, Alchemilla arvensis, Sm.

PERIWINKLE, in Chaucer and other old poets spelt PER-VINKE and PERVENKE, M.Lat. *pervincula*, dim. of L. *pervinca*, from *per*, about, and *vincire*, bind, this plant having been used for chaplets, as in the Ballad against the Scots, l. 123:

" A garlande of *pervenke* set on his heved." Ritson, vol. i. p. 33.
>> Vinca major, and minor, L.

PERSIAN WILLOW, oftener called FRENCH WILLOW, although really an American plant, from the resemblance of its leaves to willow leaves, and its foreign origin,
>> Epilobium angustifolium, L.

PERSICARIA, see PEACH-WORT.

PESTILENCE-WEED, G. *pestilenz-wurz*, the butterbur coltsfoot, from its having been formerly, as Lyte tells us, of great repute as "a sovereign medicine against the plague and pestilent fevers;" for, as the Ortus Sanitatis more explicitly declares (c. ccxlv.); " Der safft von disem kraute, gemischet mit essig und rauten-safft, yeglichs gleich vil, und dis getruncken des abents auff ein löffel foll, machet sere schwitzen, und treibet mit dem schweiss auss die pestilentz." Tussilago Petasites, L.

PETTIGREE or PETTIGRUE, Fr. *petit*, little, and *greou*, holly, the butcher's broom, so called from its prickly leaves, Ruscus aculeatus, L'Her.

PETTY-MULLEIN, the cowslip, its name in old herbals, as translated from L. *verbasculum*, this plant having been regarded as a small species of verbascum or mullein,
Primula veris, L.

PETTY-WHIN, a small prickly shrub, a name given in Lyte's Herbal to the restharrow, but by later botanists to the needle-furze, Genista anglica, L.

PEWTER-WORT, from its being used to clean pewter vessels, Equisetum hyemale, L.

PHEASANT'S EYE, from its bright red corolla and dark centre, Adonis autumnalis, L.

PICK-NEEDLE, see PINK-NEEDLE, and POWKE-NEEDLE.
Erodium moschatum, L.

PICK-PURSE, from its robbing the farmer by stealing the goodness of his land; a name that in some counties is given to the spurry, but seems to have been assigned to the shepherd's pouch more especially, on account of the number of little purses that it displays, its purse-like silicles; Capsella Bursa pastoris, L.

PIGEON'S-GRASS, Gr. περιστερεων, a place for pigeons, a name given to it, according to Galen, as quoted by Matthioli (l. iv. c. 56), from pigeons frequenting it: "quod

in ea peristeræ, hoc est columbæ, versentur." So also
the Medical MS., Sloane 1571, l. 699:

> And gyt sayth mayster Macrobius,
> Gyf yt be cast in a duffe hows,
> Alle the duffys ther abowte
> Schulle gedyr theder on a rowte.

Verbena officinalis, L.

PIGEON'S PEA, Fr. *pois-pigeon*,   Ervum Ervilia, L.

PIG-NUT, from its tubers being a favourite food of pigs,
and resembling nuts in size and flavour,

Bunium flexuosum, With.

PIG-WEED, from its being supposed to be fatal to swine,
see SOWBANE,   Chenopodium rubrum, L.

PIGGESNIE, or, as in MS. Harl. 7334, PIGGESNEYGHE, a
word that occurs in a line of Chaucer, applied to a lady,
and associated with the primrose,

> " A primerole, a *piggesnie*." C. T. 3268.

And in the ancient song, My suete swetyng, in Ritson's
collection (vol. ii. p. 21):

> "And love my pretty *pygsnye*."

The commentators on Chaucer explain it, amusingly
enough, as a "pig's eye." It seems to mean a "Whitsuntide pink," from L.G. *Pingsten*, G. *Pfingst*, and *eye*,
Fr. *oeillet*, L. *ocellus*, the name of these flowers from the
circular marking of their corolla. *Pingst* is shortened
from Gr. πεντηκοστη, meaning the fiftieth day after Easter,
and *Pinksten-eye* has been corrupted into *Piggesnie*.

Dianthus Caryophyllus, L.

PILE-WORT, L. *pila*, a ball, in allusion to the small tubers
on the roots, and its supposed efficacy, on the doctrine of
signatures, as a remedial agent,

Ranunculus Ficaria, L.

PILL-CORN, or PILD-CORN, that is, *peel-corn*, from its
grain separating from the chaff,   Avena nuda, L.

PILL-WORT, from its small globular involucres, L. *pilula*, dim. of *pila*, a ball, Pilularia globulifera, L.

PIMPINELL, or PIMPERNELL, Fr. *primprenelle*, M.Lat. *bipennella*, from having secondary little pinnæ, or feather-like leaflets; in old authors, as Evelyn in his Acetaria, and Lyte, the burnet, of which the Italian proverb says :

"L'insalata non e bella Ove non e la Pimpinella."

Poterium Sanguisorba, L.

in modern works more generally

Pimpinella Saxifraga, L.

,, RED-, a plant entirely different from the above, as are the two following species, and in no way agreeing with the name as just explained. Why it was called so, is unknown. Anagallis arvensis, L.

,, WATER-, Samolus Valerandi, L.

and in Lyte Veronica Beccabunga, and V. Anagallis, L.

,, YELLOW-, Lysimachia nemorum, L.

PIN-RUSH, Juncus effusus, L.

PINE-TREE, L. *pinus*, a word that J. Grimm considers to be a contraction of *picinus*, pitchy, and others as related to Skr. *pina*, fat, L. *pinguis*, in allusion to its resinous secretion. Isidore says that it was called so "ab acumine foliorum : *pinum* enim antiqui *acutum* nominabant" (from the sharp point of its leaves; for *picked* was by the ancients called *pinum*). The derivation of it from the Celtic *pen*, a peak, as the usual habitat of these trees, is highly improbable ; and Wedgwood's, from the English word *pin*, in respect of its pin-shaped leaves, no less so; for although this word may be an old one in the Germanic languages in the sense of "a peg," it is only within a few centuries that it has been applied to a small familiar article of female dress. Pinus, L.

,, GROUND-, from its terebinthinate odour and lowly habit, a translation of its Gr. name χαμαι-πιτυς,

Ajuga chamæpitys, L.

PINE-SAP, either from its sapping the pine, or growing from the juices of the pine, a modern term left by its author unexplained,   Monotropa Hipopitys, L.

PINK, L.Germ. *pinksten*, Whitsuntide, as in the first line of Reineke de Vos,

> "It geschach up einen pinkste dach,"
> "It happened on a Whitsunday,"

the season of flowering of one of its species, the Whitsuntide-gilliflower of old authors. The dictionaries derive it from a supposed Dutch word, *pink*, an eye; one, however, that does not appear to have any such meaning in that language. It is a curious accident, that a word, that originally meant "fiftieth," πεντηκοστη, should come to be successively the name of a festival of the Church, of a flower, of an ornament in muslin called *pinking*, of a colour, and of a sword-stab. See PIGGESNIR.   Dianthus, L.

,, CHEDDAR-, or CLIFF-, from its occurrence on Cheddar cliffs,   D. cæsius, L.

,, CLOVE-, from its odour of cloves,
          D. Caryophyllus, L.

,, CUSHION-, from its habit,   Silene acaulis, L.

,, DEPTFORD-, from its place of growth in Gerarde's time,   D. Armeria, L.

,, MAIDEN-, more properly, MEADOW-, G. *wiesennelke*, from its place of growth, a confusion of *maid* and *mead*,   D. deltoides, L.

PINK-NEEDLE, from the resemblance of its long tapering awns to the needle used in *pinking*, or making eyelet holes like pinks, in muslin,
          Erodium moschatum and cicutarium, L.

PINK-WEED, from the colour of the stems,
          Polygonum aviculare, L.

PIPE-TREE, the lilac, from its branches having a large pith that is easily bored out to make pipe-sticks, whence also its Latin name from Gr. συριγξ,   Syringa, L.

PIPEWORT,   Eriocaulon septangulare, L.

PIPPERIDGE, or PIPRAGE, red-pip, the barberry, Fr. *pepin*, a pip, and *rouge*, red, a name descriptive of the colour and character of its small juiceless fruit, which seems to be rather a pip than a berry,
Berberis vulgaris, L.

PISSABED, the dandelion, its name in nearly all the languages of Western Europe, and called so, says Dale, p. 83, " quia plus lotii derivat in vesicam, quam pueruli retinendo sunt, præsertim inter dormiendum, eoque tunc imprudentes et inviti stragula permingunt." Beckmann assigns the same reason. But it is questionable, whether a name so general could have originally belonged to a plant that has never been an article of diet, and more probable, that it has been transferred to the dandelion from the salsify or the asphodel.   Taraxacum officinale, L.

PIXIE-STOOLS, a synonym of " toad-stools " and " paddock-stools," the work of those elves,

" whose pastime
Is to make midnight mushrooms,"

and a name of some interest as showing the identity of the king of the fairies, *Puck*, with the toad, Fries. *pogge ;* for *pixie* is the feminine or diminutive of *Puck*, and the *pixie-stool* the toad-stool. The name is in the Western counties given to all suspicious mushrooms alike, but in printed books is generally assigned to the champignon.
Agaricus Chanterellus, L.

PLAISTER-CLOVER, from its trefoil leaves, and use in ointments,   Melilotus officinalis, L.

PLANE, its Old Fr. and idiomatically correct form, replaced in Spenser (F. Q. I. i. 8) and in Milton (P. L. iv. 478), with PLATANE, L. *platanus*, Gr. πλατανος,
Platanus orientalis, and occidentalis, L.

  ,,   MOCK-, the sycamore, Acer pseudoplatanus, L.

PLANTAGE, a name that occurs in Shakspeare's Troilus

and Cressida, iii. 2, with an allusion to a superstitious belief that seems to be due to a false derivation of the word, as *planet-age*, i.e. planet-wort :

"As true as steel, as *plantage* to the moon."

The poet probably meant not the plantain, but the moon-wort.   Botrychium Lunaria, L.

PLANTAIN, L. *plantago*, from *planta*, sole of the foot, and *ago*, which seems to have been used in plant-names with the sense of "wort," from the shape of the leaf in the larger species resembling a footstep,   P. major, L. etc.

,, WATER-,   Alisma Plantago, L.

PLANTAIN-SHOREWEED, a weed of the plantain tribe found beside lakes and ponds,   Littorella lacustris, L.

PLOWMAN'S ALLHEAL, see CLOWN'S-ALLHEAL.

PLOWMAN'S SPIKENARD, from the fragrant smell of the root, and its being supposed by Gerarde, p. 647, to be the Baccharis of Dioscorides, the ναρδος αγρια, *nardus rustica*, clown's nard, of other writers,   Conyza squarrosa, L.

PLUM, A.S. *plum*, L.G. *prume*, L. *prunum*, from some Asiatic name.   In Cato's time the fruit was known to the Romans, but not the tree.

Prunus communis, Huds. var. domestica, L.

POLE-REED, properly, as in Newton's Bible Herbal, POOL-REED, from its place of growth,

Arundo phragmites, L.

POLE-RUSH, properly POOL-RUSH, see BULRUSH.

POLIANTHUS, Gr. πολυς, and ανθος, many-flowered, a garden variety of the oxlip,   Primula veris, L. elatior.

POLY-MOUNTAIN, L. *polium montanum*,

Bartsia alpina, L.

POLYPODY, Gr. πολυς, and ποδες, many feet, a name given to certain ferns with pectinate fronds, from the resemblance of some of them to an insect called scolopendrium,   Polypodium, L.

POMPION, see PUMPKIN.

PONDWEED, from its growing in ponds,

Potamogeton, L.

,, HORNED-, Zannichellia, L.

POOR-MAN'S PARMACETTY, L. *sperma ceti*, whale's sperm, "the sovereignst remedy for bruises," a joke on the Latin name *Bursa*, a purse, which to a poor man is always the best remedy for his bruises ; ἀλλα συγε παντως το τραυμα ἰασαι, μικρον ἐπιπασας του χρυσιου, δεινως γαρ ἰσχαιμον ἐστι το φαρμακον. Lucian, Timon, c. 15.

Capsella Bursa pastoris, L.

POOR-MAN'S PEPPER, Lepidium latifolium, L.

POOR-MAN'S TREACLE, garlick, a translation of Lat. *Theriaca rusticorum*, the Gr. θηριακον ἀγροτων of Galen, so called because it was supposed to be an antidote to animal poisons (Plin. l. xx. c. 23). See TREACLE MUSTARD. In Arund. MS. 42, f. 15*b*, as quoted in Pr. Pm. (Way), p. 501, we read that juice of garlick "fordoþ venym and poyson mygtily, and þat is þe skyle why it is called Triacle of uppelonde, or ellys homly folkys Triacle."

Allium sativum, L.

POOR-MAN'S-WEATHER-GLASS, the red pimpernel, from its closing its flowers before rain, whence the proverb :

"No ear hath heard, no tongue can tell
The virtues of the Pimpernell."

Anagallis arvensis, L.

POPLAR, Fr. *peuplier*, from L. *popularia*, adj. of *populus*, a word that seems to be identical with *pepul*, the name of the Indian Ficus religiosa, the leaves of which so closely resemble those of the poplar, as in the varnished and pictured specimens to be very commonly taken for poplar leaves ; a name that was probably brought westward to Europe by the early Asiatic colonists, and carried eastward into India, in connexion with some religious observances,

Populus, L.

POPLAR, BLACK-, in contrast to the White poplar,.
P. nigra, L.

,,   LOMBARDY-, from a perhaps mistaken belief that it came originally from the north of Italy,
P. fastigiata, Dsf.

,,   WHITE-, or GREY-, from the colour of the under-surface of its leaves,   P. alba, L.

POPPY, A.S. *papig*, L. *papaver*,

,,   GARDEN-, or OPIUM-, or WHITE-,
P. somniferum, L.

,,   HORNED-, from its long curved seed-pods,
Glaucium luteum, L.

,,   RED-, or CORN-, or FIELD-,   P. Rhœas, L.

,,   SEA-,   Glaucium luteum, L.

,,   SPATLING-, from the froth called cuckoo-spittle so frequently found upon it,   Silene inflata, L.

,,   WELSH-, from its occurrence in Wales,
Meconopsis cambrica, L.

POTATO, Sp. *Batatas*, the name of a tropical convolvulus, the so-called "Sweet-potato," injudiciously transferred to a very different plant,   Solanum tuberosum, L.

POTHERB, WHITE-, the lamb's lettuce, in contrast to the *Olus atrum*, or Black potherb,   Valerianella olitoria, L.

POUKENEL, or POWKE-NEEDLE, L. *acus demonis*, Devil's darning needle, from *Pouke* or *Puck*, Satan, in allusion to the long beaks of its seed-vessels,   Scandix Pecten, L.

PRATLING PARNELL, a name that seems to imply a girl of suspicious character, who has let out secrets, or told tales to her own discredit.   Like the other names of this flower, London Pride, Nancy Pretty, etc., it may allude to some popular tale, song, or farce, that was in vogue in the last century.   Saxifraga umbrosa, L.

PRICKLY SAMPHIRE, see SAMPHIRE.

PRICKET, Fr. *triacquette*, dim. of *triacque*, and

PRICK-MADAM, Fr. *trique-madame*, for *triacque madame*,

from L. *theriaca,* an anthelmintic medicine, among the principal ingredients of which were stone-crops,

Sedum acre, album, and reflexum, L.

PRICK-TIMBER, or PRICK-WOOD, from its being used to make skewers, shoemakers' pegs, and goads, which were formerly called pricks, G. *pinnholtz,* the spindle tree,

Evonymus europæus, L.

PRIEST'S CROWN, from its bald receptacle, after the pappus has fallen from it, resembling the shorn heads of the Roman Catholic clergy, Taraxacum officinale, Vill.

PRIEST'S PINTLE, G. *pfaffen-pint* and *pfaffen-zagel,* Fr. *vit de prestre,* so called from the appearance of the spadix,

Arum maculatum, L.

PRIMEROLE, in Chaucer, l. 3268, from the Fr. *primeverole,* dim. of *primavera,* shortened from It. *fior di prima vera.* See PRIMROSE.

PRIMPRINT, or PRIM, a name now given to the privet, but formerly to the primrose, from the Fr. *prime printemps,* first spring, and exactly corresponding to the modern Fr. name of this flower, *primevére.* In the middle ages, however, the primrose was called in Latin *Ligustrum,* as may be seen in a Nominale of the fifteenth century in Mayer and Wright's vocabularies, p. 192 and p. 264, and several other lists, and so late as the seventeenth century in W. Coles's Adam in Eden, where he says of Ligustrum, "This herbe is called *primrose.* It is good to potage." But Ligustrum was used on the continent, and adopted by Turner, as the generic name of the Privet; and *prim-print,* as the English of Ligustrum, thus came to be transferred from the herb to the shrub. Ligustrum vulgare, L.

PRIMET, shortened from *primprint,* and correctly applied in the Grete Herball, ch. cccl., to the primrose,

Primula veris, L.

PRIMROSE, from *Pryme rolles,* the name it bears in old books and MSS. The Grete Herball, ch. cccl., says, "It is called *Pryme Rolles* of *pryme tyme,* because it beareth the first floure in *pryme tyme.*" It is also called so in Frere Randolph's catalogue. Chaucer writes it in one word *primerole.* This little common plant affords a most extraordinary example of blundering. *Primerole* is an abbreviation of Fr. *primeverole,* It. *primaverola,* dim. of *prima vera,* from *fior di prima vera,* the first spring flower. *Primerole,* as an outlandish unintelligible word, was soon familiarized into *prime rolles,* and this into *primrose.* This is explained in popular works as meaning the first rose of the spring, a name that never would have been given to a plant that in form and colour is so unlike a rose. But the rightful claimant of it, strange to say, is the daisy, which in the south of Europe is a common and conspicuous flower in early spring, while the primrose is an extremely rare one, and it is the daisy that bears the name in all the old books. See Fuchs, p. 145, where there is an excellent figure of it, titled *primula veris;* and the Ortus Sanitatis, Ed. Augsb. 1486, ch. cccxxxiii., where we have a very good woodcut of a daisy titled "masslieben, Premula veris, Latine." Brunfelsius, ed. 1531, speaking of the Herba paralysis, the cowslip, says, p. 190, expressly, "Sye würt von etlichen Doctores *Primula veris* genannt, das doch falsch ist wann *Primula veris* ist matsomen oder zeitlosen." Brunschwygk (b. ii. c. viii.) uses the same words. The Zeitlose is the daisy. Parkinson (Th. Bot. p. 531) assigns the name to both the daisy and the primrose. Matthioli (Ed. Frankf. 1586, p. 653) calls his Bellis major "*Primo fiore maggiore,* seu *Fiore di prima vera,* nonnullis *Primula veris major,*" and figures the moon-daisy. His Bellis minor, which seems to be our daisy, he calls "*Primo fiore minore, Fior di primavera,* Gallis *Marguerites,* Germanis *Masslieben.*" At p. 883 he figures the cowslip, and calls that also "*Primula veris,* Italis *Fiore di primavera,* Gallis *primevere.*" But all the older writers, as the author of the Ortus Sanitatis, Brunschwygk, Brunsfels, Fuchs, Lonicerus, and their cotemporaries, with the single exception of Ruellius, assign the name to the daisy only.

Primula veris, L. acaulis.

,, SCOTCH-, from its growth upon the mountains of Scotland,   Primula farinosa, L.

PRIMROSE PEERLESS, a name now given to a narcissus, apparently transferred to it from a lady, the favourite of Thomas à Becket, of whom it is related by Bale, that " Holye Thomas would sumtyme for his pleasure make a journey of pylgrymage to the *prymerose peerlesse* of Stafforde." Hampson, v. i. p. 121. The term *primerose* was not unfrequently applied to ladies in the middle ages. See Chaucer, C. T. l. 3268.   Narcissus biflorus, Curt.

PRINCE'S FEATHER, from its resemblance to that of the Prince of Wales,   Amarantus hypochondriacus, L.

PRIVET, in Tusser called PRIVY, altered from *Prymet*, the primrose, through a confusion between this flower and the shrub, from the application to both of them by medieval writers of the Latin *Ligustrum.*   See above PRIM-PRINT.   Ligustrum vulgare, L.

,, BARREN-, from its want of the conspicuous white flowers of the real privet, to which it certainly bears no other resemblance than in being an evergreen,

Rhamnus Alaternus, L.

PROCESSION FLOWER, see ROGATION FLOWER.

PRUNE, L. *prunea*, adj. of *prunus*, Gr. προυνη.

P. communis, Huds. var. domestica, L.

PUCKFISTS, from *fist*, G. *feist*, crepitus, and *Puck*, O.N. *puki*, who, in Pierce Plowman and other old works, seems to have been the same as Satan, but in later tales the king of the fairies, and given to coarse practical jokes. See PIXIE STOOLS.   Lycoperdon, L.

PUDDING-GRASS, pennyroyal, from its being used to make

stuffings for meat, formerly called *puddings*, as in an Old Play, The Ordinary (Dodsley vol. x. p. 229):

"Let the Corporal
Come sweating under a breast of mutton, stuffed
With *Pudding*."

R. Turner says (Bot. p. 247), that it was especially "used in Hogs-puddings," which, according to Halliwell, were made of flour, currants, and spice, and stuffed into the entrail of a hog. Mentha Pulegium, L.

PUFF-BALL, from its resemblance to a powder puff,
Lycoperdon giganteum, Bat.

PULSE, L. *puls*, Gr. πολτος, Hebr. *phul*, a pottage of meal and peas, the food of the Romans before the introduction of bread, and afterwards used to feed the sacred chicken, a term now confined to the fruit of Leguminosæ.

PUMPKIN, or POMPION, Fr. *pompon*, whence *bumpkin*, L. *pepo, -onis, Gr.* πεπων, which was used in the same sense; as, e.g. in Homer (Il. ii. 235), ὠ πεπονες, blockheads! and in the phrase πεπονος μαλακωτερος, softer than a pumpkin; see Talbot in Engl. Etym. Cucurbita Pepo, L.

PURIFICATION FLOWER, the snowdrop, see FAIR MAIDS OF FEBRUARY.

PURPLE LOOSESTRIFE, Lythrum Salicaria, L.

PURPLE MARSHWORT, or -MARSHLOCK, or PURPLE-WORT, from the colour of its flowers, and its being consequently regarded, as W. Coles tells us in his Art of Simpling, ch. xxvii., as "an excellent remedy against the purples,"
Comarum palustre, L.

PURRET, It. *porreta*, dim of *porro*, the leek, L. *porrum*,
Allium Porrum, L.

PURSLANE, in Turner PURCELLAINE, in the Grete Herball PORCELAYNE, Fr. *porcellaine*, It. *porcellana*, a name first used by Marco Polo in describing the fine earthenware made in China, and adopted from the name of a sea-shell, which resembles it in texture, and is so called from *por-*

cella, a dim. of L. *porcus* or *porca,* used in a figurative sense, as explained by Diez and Scheler. In Latin the plant was called *portulaca,* and this word seems to have been confounded with the more familiar *porcellana.* It certainly bears no resemblance to porcelain. Fuchs (Hist. plant. p. 111) derives its German name of *Portzel kraut* from L. *porcellus,* a pig. Portulaca oleracea, L.

PURSLANE, SEA-, Atriplex portulacoides, L.

PYRRIE, the pear-tree, A.S. *pirige,*

Pyrus communis, L.

QUAKERS and SHAKERS, QUAKE-, or QUAKING-GRASS, from its trembling spikelets, Briza media, L.

QUEEN OF THE MEAD, L. *Regina prati,* from its flowers resembling ostrich feathers, the badge of royalty,

Spiræa ulmaria, L.

QUICK-IN-THE-HAND, that is "Alive in the hand," the Touch-me-not, from the sudden bursting and contortion of its seed-pods upon being pressed,

Impatiens Noli me tangere, L.

QUICKEN or QUICK-BEAM, or WICKEN, a tree ever moving, A.S. *cwic-beam,* from *cwic,* alive, and *beam,* tree, translated in Ælfric's glossary "tremulus," a name applied by him to the aspen, but which has been transferred to this, the wild service, or roan tree, probably through some confusion between *cwic* and *wicce,* a witch, and the roan being regarded as a preservative against witch-craft. See ROAN. *Wicken* is merely a different spelling of the same word. *Whick* is given in Levin's Manipulus as meaning "alive," "vivus." Pyrus aucuparia, L.

QUICK-SET, from its being *set* to grow in a hedge, a *quick* or living plant, and forming what Hyll calls "a livelye hedge," as contrasted with a paling or other fence of dead wood, the hawthorn,

Cratægus Oxyacantha, L.

Quill-wort, from its resemblance to a bunch of *quills*,
                              Isoetes lacustris, L.

Quince, in Chaucer (R.R. 1. 1375) *coine*, of which *quince*
seems to be the plural, Fr. *coing*, It. and Span. *cotogna*,
L. *cydonium*, called in Greek μηλα κυδωνια, from *Cydon*, a
place in Crete,   Pyrus Cydonia, L.

Quitch-grass, or Twitch, with an interchange of the
initial consonant of frequent occurrence, owing partly,
perhaps, to the early copyists writing the letters *c* and *t*
exactly alike, but also from a dialectic tendency in some
districts to pronounce *tw* as *qu*, and *qu* as *tw*, (see Atkinson's Clev. Dial. in v. Twill) the couch-grass, A.S. *cwice*,
from *cwic*, vivacious, Sw. *kwikka*.   See Couch.
                              Triticum repens, L.

Rabone, Sp. *rabano*, L. *raphanus*, the radish,
                              Raphanus sativus, L.

Radish, It. *radice*, root, L. *radix*, a plant valued for its
root,   Raphanus sativus, L.

   „   Horse-, a larger and stronger radish,
                              Cochlearia Armoracia, L.

Ragged Robin, Fr. *Robinet dechiré*. The word *Robin*
may have reference to a popular farce of Robin and
Marion, that used to be acted in country places at Pentecost (see Ducange in v. *Robinetus*), and it is probable that
from characters in this piece the keepers' followers in the
New Forest were called *Ragged Robins*. The *Ragged* refers
to its finely laciniated petals, and seems to have suggested
the *Robin* from familiar association.
                              Lychnis Flos Cuculi, L.

Ragwort, G. *ragwurz*, a term of indecent meaning expressive of supposed aphrodisiac virtues, and originally
assigned to plants of the Orchis tribe, as it is in Germany
to the present day, and as we find it in all our own early
herbals. With the same implied meaning the pommes

d'amour are called by Lyte (b. iii. ch. 85) *Rage-apples.* In our modern floras the name *Ragwort* is, for no other assignable reason than its laciniated leaves, transferred to a large groundsel. Senecio Jacobæa, L.

RAINBERRY-THORN, (Florio in v.) the buckthorn, from L. *rhamnus.* See RHINE-BERRIES.

Rhamnus catharticus, L.

RAISIN-TREE, the red currant tree, from confusion of its fruit with the small raisins from Corinth called currants, Ribes rubrum, L.

RAMPR, in the sense of "wanton," from its supposed aphrodisiac powers, the cuckoo pint, Arum maculatum, L.

RAMPION, Fr. *raiponce,* a word mistaken, as in the cases of "cerise" and "pease," for a plural, and the *m* inserted for euphony; from L. *rapunculus,* a small *rapa,* or turnip; a bell-flower so called from its esculent tubers, Campanula Rapunculus, L.

RAMSIES or RAMSON, A.S. *hramsa,* Norw. *rams,* Da. *ramse,* Sw. *rams,* G. *ramsel,* from Da. Sw. and Ic. *ram,* rank, a wild garlick so called from its strong odour, and the rank flavour that it communicates to milk and butter. *Ramson* would be the plural of *ramse,* as peason of pease, and oxen of ox. Allium ursinum, L.

RAPE, L. *rapus,* or *rapum,* Brassica Rapa, L.

RASPBERRY, in Turner's herbal called *Raspis* or *Raspices,* of which the last syllables look like the Du. *bes, besje,* a berry. The first is more obscure. It can scarcely be *rasp,* as the dictionaries explain it; for, although the stems are rough, the fruit is not so. It seems, like several other names of plants, to be of double origin; being partly corrupted from Fr. *ronce* or *rouce,* a bramble, as *brass* from *bronce,* and partly from *resp,* as it is called in Tusser, a word that in the Eastern counties means a shoot, a sucker, a young stem, and especially the fruit-bearing stem of raspberries. (Forby.) This name it may owe to the circumstance that the fruit grows on the young shoots of the previous year. Fr. in Cotgrave *meure de ronce*.

Rubus idæus, L.

RATTLE-BOX or YELLOW RATTLE, A.S. *hrætele* or *hrætel- wyrt*, L. *crotalum*, Gr. κροταλον, from the rattling of the ripe seed in its pod, Rhinanthus Crista galli, L.

  ,, RED-, Pedicularis sylvatica, L.

RAWBONE, properly RABONE. See RUNCH.

RAY-GRASS, Fr. *ivraie*, drunkenness, from the supposed intoxicating quality of the seeds of the darnel, a species of the same genus, Lolium perenne, L.

RED-KNEES, from its red angular joints, the culrage or arsmart, Polygonum Hydropiper, L.

RED-LEGS, from its red stalks,

Polygonum Bistorta, L.

RED MOROCCO, from the colour of the petals,

Adonis autumnalis, L.

RED-ROT, from its supposed baneful effect upon sheep, and its red colour, Drosera, L.

RED-SHANKS from its red stalks,

Polygonum Persicaria, L.

and in the Northern counties, where it is a nickname for the Scotch Highlanders, the herb Robert,

Geranium Robertianum, L.

REDWEED, the red poppy, not merely from its red flowers, but from these being used as a *weed* or dye. " Ils teignent la laine en beau rouge, lorsqu'elle est traitée par l'alum et l'acide acetique." Duchesne, pl. utiles, p. 183. Papaver Rhœas, L.

REED, A.S. *hreod*, G. *riet*, and a similar name in all Germanic languages, seems to be identical with Lat. *arundo*, in which the *i* of the former is replaced with *u*, as in *hirundo* compared with Gr. χελιδων, a swallow, and an *n* inserted before *d* for euphony. The initial *h* of the A.S.

hreod is also found in the Lat. *harundo* of several MSS. and inscriptions. The root of the word unknown.

Arundo Phragmites, L.

REED-MACE, from the " Ecce homo " pictures, and familiar statues of Jesus in his crown of thorns, with this reed-like plant in his hand as a mace or sceptre,

Typha latifolia, L.

REINDEER-MOSS, a lichen on which the reindeer feeds,

Cladonia rangiferina, Hoffm.

REST-HARROW, arrest-harrow, Fr. *arrete-bœuf*, from its strong matted roots impeding the progress of plough and harrow, Ononis arvensis, L.

RHINE-BERRIES, the fruit of the buckthorn, Du. *rhyn-besien*, G. *rainbeere*, a name explained by Lyte (b. vi. c. 30) as meaning berries from the *Rhine*, by Adelung derived from *rain*, a boundary, the usual place of growth of the shrub, but perhaps from L. *rhamnus*,

Rhamnus catharticus, L.

RHUBARB, M.Lat. *rha*, from its oriental name *raved*, and *barbarum*, foreign, to distinguish this, a plant of the Volga, from the *Rha ponticum*, another kind from the Roman province, *Pontus*, Rheum, L.

„ MONK'S-, Rumex Patientia, L.

RIBBON-GRASS, the striped variety of

Digraphis arundinacea, PB.

RIBWORT, or RIBGRASS, from the strong parallel veins in its leaves, Plantago lanceolata, L.

RIE-GRASS, a name that through some confusion between *rie* and *ray* is by many farmers wrongly applied to the ray-grass, a perennial darnel, Lolium perenne, L.

but by Ray, by Martyn in his Flora Rustica, and all careful writers assigned, with more propriety, to the meadow barley, the flowering spike of which somewhat resembles that of rye, Hordeum pratense, L.

RISH, the old spelling of *rush*

ROAN-TREE, See ROWAN.

ROAST-BEEF, from the smell of the bruised leaf, the stinking gladdon,   ‑ Iris fœtidissima, L.

ROBIN-RUN-IN-THE-HEDGE, LIZZY-RUN-UP-THE-HEDGE, names of the ground ivy, which seem to have been given to it from confusion of *gill*, ferment, Fr. *guiller*, with *gill*, a girl. See GILL and HAYMAIDS. Nepeta Glechoma, B.

ROCAMBOLE, Fr. *rocambolle*, a word of uncertain derivation,   Allium Scorodoprasum, L.

ROCK-CRESS, from its alliance to the cresses, and its growth upon rocks,   Arabis stricta and petræa, Lam.

ROCK-MOSS, a lichen that grows on rocks,
                          Roccella tinctoria, Ag.

ROCK-ROSE, a name that properly belongs to the Cisti, with which the English representatives of the order were once comprised, from the resemblance of some of them to a rose, and their growth on rocks,   Helianthemum, L.

ROCK-TRIPE, Fr. *tripe de roche*, an edible lichen, upon which Sir J. Franklin and his companions supported themselves in Arctic America, and so called from some fancied resemblance,   Gyrophora vellea, Ach.

ROCKET, Fr. *roquette*, It. *rucchetta*, dim. of L. *eruca*,
                          Eruca sativa, Lam.
,, BASE-, or DYER'S-,   Reseda lutea, L.
,, BASTARD-,   Brassica Erucastrum, L.
,, DAME'S-, or GARDEN-, or WHITE-,
                          Hesperis matronalis, L.
,, LONDON-,   Sisymbrium Irio, L.
,, SEA-,   Cakile maritima, L.
,, WALL-,   Brassica muralis, Bois.
,, WINTER-, or YELLOW-, Barbarea vulgaris, RB.

ROGATION-FLOWER, from its flowering in Rogation week, the next but one before Whitsuntide, when processions were made to perambulate the parishes with the Holy Cross and Litanies, to mark the boundaries, and invoke the blessing of God on the crops. Gerarde says (p. 450) that "the maidens which use in the countries to walke the procession, make themselves garlands and nosegaies of it." It was for the same reason called Cross-, Gang-, and Procession-flower. Polygala vulgaris, L.

ROSE, L. *rosa*, a word adopted into most of the modern languages of Europe, Gr. ῥόδον, which evidently means "red," and is nearly related to Go. *rauds*, G. *roth*, W. *rhudd*, Rus. *rdeyu*, and Skr. *rohide*, red. The L. *rosa* appears to be a foreign word introduced to replace a more ancient name for this shrub, *rubus*, which, like the Gr. ῥόδον, is expressive of a red colour, as we see from its derivatives, *rubeus*, *ruber*, *rubidus*, *rubicundus*, *rubere*, *erubescere*, *rubigo*, *rubia*, but which is employed by Latin writers merely in the sense of a bramble bush. *Rosa* would seem to be connected with ῥόδον through a form in *t*, *rota*, whence *rutilus*, reddish, and L. *rota*, Wel. *rhod*, Gael. *roth*, a wheel, so named, we may presume, from the resemblance of its outline to a rose. The one cultivated in ancient times must have been a crimson species, to judge from the myth of its springing from the blood of Adonis; Homer's ῥοδοδάκτυλος Ἡώς; the comparison of it with Tyrian purple in Columella's line:

Jam rosa mitescit Sarrano clarior ostro;

and the distinct statement of Isidore (c. ix.) that it was called so, "quod rutilante colore rubet." Rosa, L.

,, BRIER-, or DOG-, R. canina, L.

,, BURNET-, from the resemblance of its leaf to that of the burnet, R. spinosissima, L.

,, CANKER-, from its supposed injurious effect on wheat-crops, the red or field poppy, Papaver Rhœas, L.

,, CHRISTMAS-, from its rose-like flowers, and its blossoming in the winter, Helleborus niger, L.

,, CORN-, the field poppy, Papaver Rhœas, L.

,, GUELDER-, from its balls of white flowers which

somewhat resemble a double rose, and its native country Gueldres, the sterile-flowered var. of the water elder,

Viburnum Opulus, L.

Rose, Province-, from *Provins*, a small village near Paris, where it used to be cultivated, R. gallica, L. var.

Rose-a-ruby, L. *rosa rubea*, from its rich red flowers,

Adonis autumnalis, L.

Rose Bay, the name given by Turner to the oleander, but now, from resemblance of leaf in an outline drawing, applied in some books to a very different plant,

Epilobium angustifolium, L.

Rose Campion, the rose-coloured campion,

Lychnis coronaria, L.

Rose Elder, the elder that bears roses, the Guelder rose, Viburnum Opulus, L.

Rose-root, or -wort, L. *rhodia radix*, from the odour of its rootstock, Rhodiola rosea, L.

Rosemary, L. *rosmarinus*, sea-spray, from its usually growing on the sea-coast, and its odour,

Rosmarinus officinalis, L.

,, Marsh-, or Wild-, from its narrow linear leaves like those of rosemary, Andromeda polifolia, L.

Rot-grass, from its being supposed to bane sheep, a *grass* in the sense of herbage, Pinguicula vulgaris, L.

Rowan, or Roan-tree, called in the Northern counties Ran or Royne, Da. and Sw. *rönn* or *runn*, the O.Norse *runa*, a charm, from its being supposed to have power to avert the evil eye. "The most approved charm against cantrips and spells was a branch of the *Rowan-tree* planted and placed over the byre. This sacred tree cannot be removed by unholy fingers." Jamieson's Scot. Dict[y].

> "*Roan-tree* and red thread
> Haud the witches a' in dread."

Johnston in East. Bord.

The word *runn*, from Skr. *ru*, murmur, meant a secret.

A *rûn-wita* was a private secretary, one who knew his master's secrets ; and from the same word were derived *rynan*, to whisper, *rûna*, a whisperer, in earlier times a magician, and *rûn-stafas*, mysterious staves. From this last use of the word the name *run* came naturally to be applied to the tree from which such staves were usually cut, as *boc* to that from which bookstaves, *bocstafas*, were made ; but it does not appear to be ascertained why this tree should have been so exclusively used for carving runes upon, as to have derived its name from them, not only in the British isles, but in the Scandinavian countries also. There was probably a superstitious feeling of respect for it derived from ancient times.

Pyrus Aucuparia, Gärt.

RUDDES, a name that should mean a red or ruddy flower, and is hardly applicable in the present sense of the word to a yellow one, such as the marigold, to which it is given in early writers. But *ruddy* was formerly said of gold ; and the author of the Grete Herball, in speaking of this plant, says, " Maydens make garlands of it, when they go to feestes and bryde ales, because it hath fayre yellowe floures and ruddy." Calendula officinalis, L.
and also Chrysanthemum segetüm, L.

RUE, L. *ruta*, its meaning unknown,

Ruta graveolens, L.

,, MEADOW-, or FEN-, from its rue-like much divided leaves, and its place of growth, Thalictrum flavum, L.

,, WALL-, Asplenium Ruta muraria, L.

RUNCH, a word that in Scotland is applied to a strong rawboned woman, as a " *runchie* quean," in reference, as Jamieson thinks, to a coarse wild radish, the jointed charlock, so called from another meaning of the word *runch*, viz., to crunch, Raphanus Raphanistrum, L.

RUPTURE-WORT, from its fancied remedial powers,

Herniaria glabra, L.

RUSH, called in old authors RYSCHYS, RISH, RESH, and RASHES, A.S. *risc*, related to It. *lisca*, reed, Go. *raus*,

		Juncus, L.
,,	BOG-,	Schœnus, L.
,,	BUL-,	Scirpus lacustris, L.
,,	CLUB-,	Scirpus palustris, L.
,,	DUTCH-, or SCOURING-,	Equisetum hyemale, L.

,, FLOWERING-, from its tall rush-like stem and handsome head of flowers,   Butomus umbellatus, L.

,, PIN-,   Juncus effusus, L.

RUST, from an effect similar to the rust of iron produced upon plants by certain minute fungi.

RUTABAGA, the Swede turnip, so called from Sw. *rotabaggar*, root-rams.   J. H. Lundgren in N. & Q., 4th ser., v. 76.   Brassica campestris, L. var. rutabaga.

RYE, A.S. *ryge*, O.N. *rugr*, W. *rhyg*, O.H.G. *roggo*, Lith. *ruggei*, Rus. *rosh*, Pol. *rez'*, Esth. *rukki*, a word extending, with dialectic modifications, all over Northern Asia, from which this grain seems to have travelled to the South and West. Its derivation unknown. See L. Diefenbach, Or. Eur. No. 29, and J. Grimm, Gesch. d. D. Spr. p. 64.

Secale cereale, L.

RYE-GRASS, see RIE-GRASS, and RAY-GRASS.

SABIN, see SAVINE.

SAFFLOWER, from its flowers being sold, as a dye, for genuine saffron,   Carthamus tinctorum, L.

SAFFRON, Sp. *azafran*, Ar. *al zahafaran*,

Crocus sativus, L.

SAGE, Fr. *sauge*, It. and Lat. *salvia*, which by change of *l* to *u* became *sauuia*, *sauja*, *sauge*, as *alveus*, a trough, by the same process, *auge*,   Salvia, L.

,, WOOD-, from its sage-like leaves, and growth in woods, and about their borders, Teucrium Scorodonia, L.

SAINFOIN, sometimes spelt, as in Lyte, in Dale, and in

Martyn's Flora Rustica, *Saintfoin*, in Hudson *St. Foin*, in Plukenet *Sainct-foin*, and thence translated by some of our old writers " Holy hay," but really formed from Fr. *sain*, wholesome, and *foin*, hay, L. *sanum fœnum*, representing its older name *Medica*, which properly meant " of Media," but was misunderstood as meaning "curative." According to Plukenet and Hill, the name was first given to the lucerne, Medicago sativa, and that of lucerne to an Onobrychis, our present sainfoin. There does not appear to be any saint named *Foin*, nor any reason for ascribing divine properties to this plant. According to Bomare quoted by Duchesne, "Le S. est ainsi appelé parceque c'est le fourage le plus appetisant, le plus nourrissant, et le plus *sain*, qu'on puisse donner aux chevaux et aux bestiaux." Good reasons for a name follow of course. The equivocal word *Medica* is undoubtedly the origin of this one. See MEDICK. As at present applied, Onobrychis sativa, Lam.

SAINFOIN, in the Dictionary of Husbandry, 1717, lucerne, which is explained as " Medick fodder, Spanish trefoil, and Snail or Horned clover grass," Medicago sativa, L.

ST. ANTHONY'S NUT, the pignut, from his being the patron saint of pigs. " Immundissimas porcorum greges custodire cogitur miser Antonius." Moresini Papatus, p. 133. Bunium flexuosum, L.

ST. ANTHONY'S RAPE or TURNEP, from its tubers being a favourite food of pigs, Ranunculus bulbosus, L.

ST. BARBARA'S CRESS, from its growing in the winter, her day being the 4th Dec. old style; or as the Grimms explain the G. synonym *barbel-* or *barben-kraut*, " weil es die *barben* im bach fressen," because the barbel in the brook eat it, Barbarea vulgaris, DC.

ST. BARNABY'S THISTLE, from its flowering at the summer solstice, the 11th June, old style, now the 22nd, his day, whence its Latin specific name,

Centaurea solstitialis, L.

St. Bennet's herb, see Herb Bennet.

St. Catharine's flower, from its persistent styles resembling the spokes of her wheel, Nigella damascena, L.

St. Christopher's herb, see Herb Christopher.

St. Daboec's heath, from an Irish saint of that name, a species found in Ireland,　　　Menziesia polifolia, Jus.

St. James's wort, either from its being used for the diseases of horses, of which this great warrior and pilgrim saint was the patron; or, according to Tabernæmontanus, because it blossoms about his day, the 25th July, which may have led to its use in a veterinary practice upon male colts at this season;　　　Senecio Jacobæa, L.

St. John's wort, from its being gathered on the eve of St. John's day, the 24th June, to be hung up at windows as a preservative against thunder and evil spirits, whence it was called Fuga dæmonum, and given internally against mania,　　　Hypericum perforatum, L.

St. Patrick's Cabbage, from its occurrence in the West of Ireland, where St. Patrick lived, the London pride,
　　　　　　　Saxifraga umbrosa, L.

St. Peter's wort, of the old Herbals, the cowslip, from its resemblance to St. Peter's badge, a bunch of keys, whence G. *schlüssel-blume*,　　　Primula veris, L.

St. Peter's wort of modern floras, from its flowering on his day, the 29th June,　Hypericum quadrangulum, L.

Salad, or Sallet, Corn-, from it being eaten as a salad, and growing in corn-fields,　Valerianella olitoria, Poll.

Salad Burnet, a burnet eaten with salad,
　　　　　　　Poterium sanguisorba, L.

Salep, Mod. Gr. σαλεπι, Pers. *sahaleb*, the plant from which salep is made,　　　Orchis latifolia, L.

Saligot, Fr. *saligot*, a sloven, or one who lives in dirt, from its growing in mud, a plant that Lyte tells us (p. 536) was found in his time " in certayne places of this countrie, as in stues and pondes of cleare water," Trapa natans, L.

SALLOW, A.S. *sealh, salh, salig,* O.H.G. *salaha,* Da. *selje,* O.N. *selja,* L. *salix,* Gr. ἑλιξ, Ir. *sail* and *saileog,* Sw. *salg,* Fin. *salawa,* different forms of a word that implies a shrub fit for withes, A.S. *sal,* or *sœl,* a strap or tie, with a terminating adjectival *ig* or *h,* corresponding to the *ix,* or *ex,* or *ica* in the Latin names of shrubs. *Sal,* a hall, in O.H.G. a house, G. *saal,* seems to be of the same origin, and to tell us that our ancestors dwelt in houses of wicker work, even men of rank. The L. *aula,* Gr. ἀυλη, is perhaps the same word as *sal.* It means both a stall and a hall. In fact, the royal sheepcote was in the primitive nation the royal palace, as among the Tartars of the interior of Asia is the *aoul* at the present day. See Westmacott, p. 84.

Salix, L.

SALLOW-THORN, from its white willow-like leaves, and spinous branches, Hippophae rhamnoides, L.

SALSIFY, Fr. *salsifis,* L. *solsequium,* from *sol,* sun, and *sequi,* follow, a plant whose flower was supposed to follow the sun, Tragopogon porrifolius, L.

SALTWORT, from its officinal Latin name *Salicornia,* salthorn, Salicornia herbacea, L.

   ,, BLACK-, Glaux maritima, L.

SAMPHIRE, more properly spelt SAMPIRE or SAMPIER, Fr. *Saint Pierre,* It. Herba di *San Pietro,* contracted to *Sampetra;* from being, from its love of sea-cliffs, dedicated to the fisherman saint, whose name is the Gr. πετρος, a rock, Fr. *pierre,* Crithmum maritimum, L.

   ,, MARSH-, Salicornia herbacea, L.

SAND-WEED, or -WORT, from its place of growth, Arenaria, L.

SANGUINARY, L. *sanguinaria,* the yarrow, from being confused, under the equivocal name *millefolium,* with a horsetail that Isidore tells us (c. ix.) was formerly used to make the nose bleed, and thence called *herba sanguinaria.* See NOSEBLEED. Achillæa Millefolium, L.

SANICLE, a word usually derived immediately from L. *sanare*, heal, which on principles of etymology is impossible. Indeed it is, as Adelung remarks, an even question, whether its origin is Latin or German. Its great abundance in the middle and north of Europe would incline us rather to the latter as the likeliest, and it may be a corruption of *Saint Nicolas* called in German *Nickel*. Whatever its derivation, the name was understood in the Middle ages as meaning "curative," and suggested many proverbial axioms, such as:

"Qui a la bugle et la *sanicle*, Fait aux chirurgions la nicle."

" He who has bugle and sanicle makes a joke of the surgeons ; " and

" Celuy qui *sanicle* a, De mire affaire il n'a."

" He who keeps sanicle, has no business with a doctor." *Sanicula* does not occur in classical Latin writers, and there is no such word as *sanis* or *sanicus* from which it could have been formed. But in favour of the derivation from *San Nicola* or *Sanct Nickel*, is the wonderful Tale of a Tub, the legend of his having interceded with God in favour of the two children, whom an innkeeper had murdered and pickled in a pork tub, and obtained their restoration to life and health. See Forster's Perennial Calendar, p. 688, and Mrs. Jameson's Sacred and Legendary Art, p. 273. A plant named after this saint, and dedicated to him, might very reasonably be expected to "make whole and sound all wounds and hurts both inward and outward," as Lyte and other herbalists tell us of the sanicle. The Latin name, as in so many other cases, would be the nearest approach that could be made to the German. See SELF-HEAL. Sanicula europœa, L.

SARACENS CONSOUND, M.Lat. *Consolida Saracennica*. Parkinson says (Th. Bot. p. 540), that "it is called *Solidago* and *Consolida* from the old Latine word *consolidare*, which in the barbarous Latine age did signify to soder, close, or glue up the lips of wounds; and *Saracenica,* because the Turks and Saracens had a great opinion thereof in healing the wounds and hurts of their people, and were accounted great chirurgions, and of wonderful skill therein." Hence it was in German also called *Heidnisch wundkraut.*

Senecio Saracenicus, W.

SATIN-FLOWER, from the satiny dissepiments of its seed vessel, Lunaria biennis, L.

SATYRION, L. *satyrium,* Gr. σατυριον from σατυρος, a satyr, a name applied to several species of orchis, from their supposed aphrodisiac character. "Mulieres partium Italiæ dant eam radicem tritam cum lacte caprino ad incitandam libidinem." Herbarius, c. cxxviii. Orchis, L.

SAUCE-ALONE, so called, according to W. Coles, from being "eaten in spring-time with meat, and so highly flavoured that it serves of itself for sauce instead of many others." This is an ingenious explanation of the name; but the real origin of it is more likely to be the It. *aglione,* Fr. *ailloignon,* coarse garlick. Like its German name, *Sasskraut,* sauce-herb, the English will mean "sauce-garlick," and refer to its strong alliaceous odour.

Erysimum Alliaria, L.

SAUGH, the sallow, A.S. *sealh,* Salix Caprea, L.

SAVINE, from the *Sabine* district of Italy, Juniperus sabina, L.

SAVOURY, Fr. *savorée,* It. *savoreggia,* L. *satureja,* Satureja hortensis, L.

SAVOY, from the country of its discovery, Brassica oleracea, L. var. Sabauda.

SAW-WORT, from its leaves being nicked like a saw, Serratula arvensis, L.

SAXIFRAGE, L. *saxifraga,* from *saxum,* rock, and *frango,* break, being supposed to disintegrate the rocks, in the crevices of which it grows, and thence, on the doctrine of signatures, to dissolve stone in the bladder. Isidore of

Seville derives it primarily from this latter quality. The
words in the Ort. San. are: " Der meister Ysidorus spricht,
das dises kraut umb des willen heysst saxifraga, wann es
den stein brichet in der blasen, und den zu sandt machet."
It is for the same reason called in Scotland *Thirlstane.*

Saxifraga, L.

also from its supposed similar virtues,

Pimpinella Saxifraga, L.

SAXIFRAGE, BURNET-,	Pimpinella Saxifraga, L.
„ GOLDEN-,	Chrysosplenium, L.
„ MEADOW-,	Silaus pratensis, Bess.

SCABWORT, from its use in veterinary medicine to cure
scabby heels, the elecampane,   Inula Helenium, L.

SCABIOUS, L. *scabiosa,* scurfy, from *scabies,* scurf, in allusion to the scaly pappus of its seeds, which, on the doctrine
of signatures, led to its use in leprous diseases, and its
being regarded as a specific remedy for all such as were
"raüdig" or "grindig," itchy or mangy. See Brunschwygk.

Scabiosa, L.

SCAD-TREE, in Jacob's Pl. Faversh. the bullace,

Prunus insititia, Huds.

SCALD-BERRY, from the supposed curative effect of its
leaves boiled in lye in cases of *scalled* head, Park. Th. Bot.,
p. 1016, the blackberry,   Rubus fruticosus, L.

SCALE-FERN, from the scales that clothe the back of the
fronds,   Ceterach officinarum, W.

SCALLION, a garlick from *Ascalon* in Syria,

Allium ascalonicum, L.

SCARLET-RUNNER, a climber with scarlet flowers,

Phaseolus multiflorus, L.

SCIATICA-CRESS, from Lat. *Ischiatica,* so called from its
supposed effect in cases of irritation of the ischiatic nerve,
a species of candytuft,   Iberis amara, L.

SCORPION-GRASS, the old name of the plant now called
" Forget-me-not," and that under which it is described

in all our Herbals, and all our Floras, inclusive of the
Flora Londinensis and Gray's Natural Arrangement, till
the end of the first quarter of this century, when the term
"Forget-me-not" was introduced with a pretty popular
tale from Germany, and superseded it. It was called
Scorpion-grass from being supposed, on the doctrine of
signatures, from its spike resembling a scorpion's tail, to be
good against the sting of a scorpion. Lyte tells us (b. i.
c. 42) that in his day, 1578, it had "none other knowen
name than this." Myosotis, L.

SCOTCH ASPHODEL, a plant of the Asphodel tribe common in Scotland, Tofieldia palustris, Huds.

SCOTCH FIR, from its growing wild in Scotland,
Pinus sylvestris, L.

SCOTCH THISTLE, the thistle adopted as the badge of
Scotland in the national arms, usually taken to be the
musk thistle, Carduus nutans, L.
but according to Johnston in East. Bord.
Onopordon Acanthium, L.

SCOURING RUSH, or SCRUB-GRASS, a rush-like plant used
in scouring utensils of wood or pewter, the Dutch rush, a
species of horsetail, Equisetum hyemale, L.

SCRAMBLING ROCKET, a corruption of *Crambling.*

SCRATCHWEED, Fr. *grateron,* from *gratter,* scratch, the
goose-grass or cleavers, Galium Aparine, L.

SCURVY-GRASS, -CRESS, or -WEED, from its use against
scurvy, Cochlearia officinalis, L.

SEA-BEET, Beta maritima, L.

SEA-BELLS, -BINDWEED, or -WITHWIND,
Convolvulus Soldanella, L.

SEA-BELT, Laminaria saccharina, Lam.

SEA-BUCKTHORN, Hippophae rhamnoides, L.

SEA-BUGLOSS, Pulmonaria maritima, L.

SEA-CALE, or -KALE, a cale or colewort that grows by
the sea-side. Crambe maritima, L.

SEA-GILLIFLOWER,	Statice Armeria, L.
SEA-GRAPE,	Salicornia herbacea, L.
SEA-GRASS,	Ruppia maritima, L.
SEA-HARD-GRASS,	Ophiurus incurvatus, RB.
SEA-HEATH,	Frankenia pulverulenta, L.
SEA-HOLLY, -HOLME, or -HULVER,	
	Eryngium maritimum, L.
SEA-LACES, -CATGUT, or -WHIPCORD,	
	Chorda Filum, Lam.
SEA-LAVENDER,	Statice Limonium, L.
SEA-LYME-GRASS,	Elymus arenarius, L.
SEA-MAT-WEED,	Psamma arenaria, P.B.
SEA-MILK-WORT,	Glaux maritima, L.
SEA-POPPY,	Glaucium luteum, L.
SEA-PURSLANE,	Atriplex portulacoides, L.
SEA-REED,	Psamma arenaria, P.B.
SEA-ROCKET,	Cakile maritima, L.
SEA-STARWORT,	Aster Tripolium, L.
SEA-WEEDS,	Algæ.

SEAL-WORT, from the round markings, like impressions of a seal, on the root-stock, the Solomon's seal,

Convallaria Polygonatum, L.

SEAVES, rushes, a North-country word, Da. *siv*,

Juncus, L.

SEDGE, SEGG, or SEGS, originally the same word, A.S. *secg*, which is identical with *sæcg*, and *seax*, a small sword, a dagger, and was applied indiscriminately to all sharp-pointed plants growing in fens, rushes, reeds, and sedges. Thus in a Wycliffite version of Exod. ii. 3, "she took a basket of rush," is, "sche took a leep of *segg*." Their sense is at present limited; *Sedge* being now confined to the genus Carex, L. and *Segg* to the gladdon and flag-flowers, Iris, L.

SEE-BRIGHT, from its supposed effect on the eyes (see CLARY), Salvia Sclarea, L.

SEGGRUM, from its application as a vulnerary to newly-cut rams, bulls, and colts, which in the North are called *seggrams* and *seggs*. See STAGGERWORT.

. Senecio Jacobæa, L.

SELF-HEAL, correctly so spelt, and not *Slough-heal*, for reasons stated under this latter term. It meant that with which one may cure one's self, without the help of a surgeon, to which effect Ruellius quotes a French proverb, that " No one wants a surgeon who keeps Prunelle." See Park. Th. Bot. p. 526. Prunella vulgaris, L. and also, for the same reason, Sanicula europæa, L.

SENGREEN, A.S. *sin*, ever, and *grene*, green, from its evergreen leaves, the houseleek,

Sempervivum tectorum, L.

SENVY, Fr. *sénevé*, G. *senf*, L. *sinapis*, Gr. σιναπι, mustard, Brassica nigra, Boiss.

SEPTFOIL, or SETFOIL, from its seven leaflets, Fr. *sept*, and *feuilles*, Lat. in Apuleius, (c. 117,) *septefolium*, Gr. ἑπταφυλλον, Potentilla Tormentilla, Sib.

SERVICE-, or, as in Ph. Holland's Pliny more correctly spelt, SERVISE-TREE, from L. *cervisia*, its fruit having from ancient times been used for making a fermented liquor, a kind of beer :

" Et pocula læti
Fermento atque acidis imitantur vitea *sorbis*."
Virg. Geor. iii. 379.

Diefenbach remarks : (Or. Eur. 102) " bisweilen bedeutet cervisia einen nicht aus Getreide gebrauten Trank ; " and Evelyn tells us in his Sylva (ch. xv), that " ale and beer brewed with these berries, being ripe, is an incomparable drink." The *Cerevisia* of the ancients was made from malt, and took its name, we are told by Isidore of Seville, from *Ceres, Cereris*, but this has come to be used in a secondary sense without regard to its etymological meaning, just as in *Balm-tea* we use *tea* in the sense of an

14

infusion, without regard to its being properly the name of
a different plant. Pyrus domestica, Sm.

SERVICE, WILD-, the rowan tree, Pyrus aucuparia, Gärt.

SETWALL, from M.Lat. *Zedoar* or *Zeduar*, the name of
an Oriental plant for which this was sold, through the
changes of *r* to *l*, and *z* to *s*, by which we get *Zeduar*,
Zedualle, *Setewale*, as in Chaucer, and *Setwal*, a plant
usually understood to be Valeriana officinalis, L.
but according to Lyte Valeriana pyrenaica, L.

SETTERWORT, a plant so called because it was used for
the operation of *settering*. " Husbandmen are used to
make a hole, and put a piece of the root into the dewlap
of their cattle, as a *seton*, in cases of diseased lungs ; and
this is called pegging or *settering*." (Gerarde, p. 979.)
Setter is a corruption of *seton*, It. *setone*, a large *seta*, or
thread of silk. Helleborus foetidus, L.

SHAKER, from the tremulous motion of its spikelets, a
synonym of its other name, *quaker*, Briza media, L.

SHALLOT, Fr. *eschalotte*, from L. *Ascalonitis*, of *Ascalon*
in Palestine, its native country, Allium ascalonicum, L.

SHAMROCK, Erse *seamrog*, compounded of *seamar*, trefoil,
and *og*, little, the *seamar*, in the opinion of L. Diefenbach,
the same as *sumar*, in a word that is given by Marcellus
of Bordeaux, physician to Theodosius the Great, as the
Celtic name of clover, *visumarus*. The plants that for a
long time past have been regarded by the Irish as the true
shamrock, and worn by them on St. Patrick's Day, are the
black nonsuch, and the Dutch clover ; and these, but
chiefly and almost exclusively the first, are sold for the
national badge in Covent Garden, as well as in Dublin.
Intermixed with them are several other species of the same
two genera, medicago and trifolium, but no plant of any
other genus. Of late years, however, several writers have
adopted Mr. Bicheno's fancy, and advocated the claims of
the wood-sorrel to this honour, but certainly without the

smallest shadow of reason. Mr. J. Hardy, in an excellent article on the subject in the third number of the Border Magazine, has shown that the plant intended by the writers of Queen Elizabeth's reign was the watercress. Thus Stanihurst, in Holinshed's Chronicle, ed. 1586, says : " Watercresses, which they tearme *shamrocks*, roots, and other herbs they feed upon ;" a statement which he repeats in his work, " De rebus in Hibernia," p. 52. Fynes Morison also says that "they willingly eate the herbe shamrock, being of a sharp taste, which, as they run to and fro, they snatch like beasts out of the ditches." It will be objected to the watercress, that its leaf is not trifoliate, and could not have been used by St. Patrick to illustrate the doctrine of the Trinity. But this story is of modern date, and not to be found in any of the lives of that saint. In Chambers' Book of Days, vol. i. p. 384, it is stated that the trefoil is in Arabic called *shamrakh*, and held sacred in Iran, as emblematical of the Persian Triads. The word is *shimrakh*, and means a bunch of dates ! The plant which is figured upon our coins, both Irish and English, is a conventional trefoil. As its leaflets are stipitate, it cannot have been meant for a wood-sorrel, as some writers have pretended it to be. The plant that, upon the authority of Dr. Moore, of Dublin, and other competent persons, has for many years been recognized in Ireland as the true sham·rock, is the black nonsuch, Medicago lupulina, L. and occasionally mixed with it, or mistaken for it, the Dutch clover, Trifolium repens, L.

SHAREWORT, L. *inguinalis*, from being supposed to cure diseases of the share or groin, called *buboes*, whence one of its synonyms in old authors *bubonium*. The misunderstanding of this word *bubonium* led to some ludicrous theories of the effect of the plant upon toads. The Ortus Sanitatis tells us (ch. 431) that it means *toad-wort*, for that " *bubo* means *toad*. Inde *bubonium*. And it is so called, because it is a great remedy for the toads. When a spider stings a toad, and the toad is becoming vanquished, and the spider stings it thickly and frequently, and the toad cannot avenge itself, it bursts asunder. But if such a burst toad is near this plant, it chews it, and becomes sound again. But if it happens that the wounded toad cannot get to the plant, another toad fetches it, and gives it to the wounded one." A case is recorded in Topsell's Natural History, p. 729, as having been actually witnessed by the Duke of Bedford and his attendants, at a place called Owbourn, (a mis-spelling perhaps of Woburn,) and oftentimes related by himself. The error has arisen from the confusion of *bubo* with *bufo*. The toad-flax has acquired its name from a similar blunder. Aster Tripolium, L.

SHAVE-GRASS, from being "used by fletchers and comb-makers to polish their work therewith," says W. Coles, Du. *schaaf-stroo*, from *schaaf*, a plane, and *stroo*, straw,

Equisetum hyemale, L.

SHEEP'S-BANE, from its character of baning sheep (see Ger. p. 528), the whiterot, Hydrocotyle vulgaris, L.

SHEEP'S-BIT, or SHEEP'S-BIT-SCABIOUS, so called to distinguish it from the Devil's-bit-scabious,

Jasione montana, L.

SHEEP'S PARSLEY, in Suffolk,

Chærophyllum temulum, L.

SHEEP'S SORREL, Rumex Acetosella, L.

SHELLEY GRASS, or, as Threlkeld spells it, SKALLY GRASS, a word the origin of which is obscure, perhaps the Sc. *skellie*, which means "charlock," extended to weeds in general, the couch-grass, Triticum repens, L.

SHEPHERD'S CRESS, Teesdalia nudicaulis, RB.

SHEPHERD'S NEEDLE, Scandix Pecten, L.

SHEPHERD'S PURSE, Capsella Bursa, DC.

SHEPHERD'S ROD, or STAFF, L. *virga pastoris*,

Dipsacus pilosus, L.

SHEPHERD'S, or POOR MAN'S WEATHER-GLASS, from its closing its flowers before rain, the pimpernel,

Anagallis arvensis, L.

SHERE-GRASS (Turn. i. 112), sedge, from its cutting edges, A.S. *sceran*, shear, Carex, L.

SHORE-GRASS, or SHORE-WEED, from its usual place of growth, Littorella lacustris, L.

SICKLE-WORT, L. *secula*, from the shape of its flowers, which seen in profile resemble a sickle,

Prunella vulgaris, L.

SIETHES, in Tusser, a kind of chives, spelt in Holybande SIEVES, from the Fr. *cive*, Allium fissile, L.

SILVER FIR, from its silvery whiteness,

Pinus picea, L.

SILVER-WEED, L. *Argentina*, from the silvery glitter of the under surface of its leaves, Potentilla anserina, L.

SIMPLERS' JOY, from the good sale they had for so highly esteemed a plant, the vervain, Verbena officinalis, L.

SIMSON, Fr. *seneçon*, in Bulleyn SENTION, and in the Eastern counties SENCION, corruptions of L. *senecio, -onis*, a name derived from *senex*, an old man, and given to the common groundsel in allusion to its heads of white hair, the pappus upon the seed:

"Quod canis similis videatur flore capillis."—Macer;

or, as Bulleyn expresses it: "because the flower of this herbe hath white hair, and when the winde bloweth it away, then it appereth like a bald-headed man."

Senecio vulgaris, L.

SINKFIELD, a corruption of *cinquefoil*, Potentilla, L.

SKEG, the sloe-tree, in Ph. Holland's Pliny (b. xviii. c. 6) and Florio, from its rending clothes, as a *sceg*, or ragged projecting stump might, Prunus spinosa, L.

SKEWER-WOOD, from skewers being made of it, the spindle-tree, Evonymus europæus, L.

SKIRRET, in old works called SKYRWORT or SKYRWYT, Du. *suikerwortel*, sugar-root, Sium Sisarum, L.

SKULLCAP, from the shape of the calyx,

Scutellaria galericulata, L.

SLEEP-AT-NOON, from its flowers closing at midday, the goat's beard, Tragopogon pratense, L.

SLEEPWORT, from its narcotic properties,

Lactuca virosa, L.

SLOE, in Lancashire *slaigh* or *sleawgh*, A.S. *sla-*, *slag-*, or *slah-þorn*, the *sla* meaning not the fruit, but the hard trunk, a word that we find in our own, and in all its kindred languages, to be intimately connected with a verb meaning *slay* or strike.

NOUN.	VERB.	NOUN.	VERB.
A.S. sla	slean	Da. slaaen	slaa
slage	slagan	Sw. slå	slå
Eng. sloe, O.E. sle	slay	Icel.	sla
Du. and L.G. slee	slaan	Old Fries.	sla
G. schlehe	schlagen	Old Sax.	slahan or slan

Whether this connexion is due to the wood having been used as a flail (as, from its being so used at this day, is most probable) or as a bludgeon, can only be discovered by a comparison of its synonyms and the corresponding verb in other languages of the Ind-European group.

Prunus spinosa, L.

SLOKE, or SLAKE, a name given to several species of edible Porphyræ and Ulvæ.

SLOUGH-HEAL, a supposed, but mistaken correction of *Self-heal*, the *slough* being that which is thrown off from a foul sore, and not that which is healed, the wound itself. Besides, the term *slough* was not used in surgical language till long after the plant had been called *Selfe-heal*, and applied as a remedy, not to sloughing sores, but to fresh cut wounds. See SELF-HEAL. Prunella vulgaris, L.

SMALLAGE, a former name of the celery, meaning the

small ache or parsley, as compared with the ἱπποσελινον, or great parsley, olus atrum. See Turner's Nomenclator, A.D. 1548, and Gerarde. See also ACHE.

Apium graveolens, L.

SMOKE-WOOD, from boys smoking its porous stalks,

Clematis Vitalba, L.

SMUT, from its resemblance to the smut on kettles,

Uredo caries, L.

SNAG, in Cotgrave, and in Lyte (b. vi. ch. 47), the sloe, from its branches being full of small snags or projections,

Prunus spinosa, L.

SNAIL CLOVER, from the spiral convolutions of its legumes, the lucerne genus,　　　　　　Medicago, L.

SNAKE'S HEAD, from the checkered markings on the petals like the scales on a snake's head,

Fritillaria Meleagris, L.

SNAKE'S TAIL, from its cylindrical spikes, the sea hard-grass,　　　　　　Ophiurus incurvatus, RB.

SNAKE-WEED, the bistort, from its writhed roots,

Polygonum Bistorta, L.

SNAP-DRAGON, from its corolla resembling the *snap* or snout, Du. *sneb*, G. *schnabel*, of some animal. It means, perhaps, "Snap, dragon!"　　　Antirrhinum majus, L.

SNEEZE-WORT, from the powder made from it causing to sneeze, L. *sternutamentoria, Gr.* πταρμικη,

Achillæa Ptarmica, L.

SNOW-BALL TREE, from its round balls of white flowers, the Guelder rose, a cultivated variety of the water-elder,

Viburnum Opulus, L.

SNOWBERRY, from the white colour and snowlike pulp of its fruit,　　　　　Symphoria racemosa, Ph.

SNOW-DROP, from G. *schneetropfen*, a word that in its usually accepted sense of a drop of snow is inconsistent; for a dry powdery substance, like snow, cannot form a drop. In fact, the *drop* refers not to icicles, but to the large pendants, or drops, that were worn by the ladies in the sixteenth and seventeenth centuries, both as ear-rings and hangings to their brooches, and which we see so often represented by the Dutch and Italian painters of that period. Galanthus nivalis, L.

SNOW-FLAKE, a name invented by W. Curtis to distinguish it from the snow-drop, Leucojum æstivum, L.

SOAPWORT, from its being used in scouring (Ger. p. 360), and frothing in the hands like soap, says Brunschwygk,

Saponaria officinalis, L.

SOLDIER-ORCHIS, from a fancied resemblance in it to a soldier, Orchis militaris, L.

SOLOMON'S SEAL, from the flat round scars on the rootstock, resembling what is called a Solomon's seal, a name given by the Arabs to a six-pointed star, formed by two equilateral triangles intersecting each other, and of frequent occurrence in Oriental tales,

Convallaria Polygonatum, L.

SOPS-IN-WINE, from the flowers being used to flavour wine. Chaucer says of it, writing in Edw. III's reign :

"There springen herbes grete and smal,
The licoris and setewale,
And many a *clove gilofre*,
To put in ale,
Whether it be moist or stale." C.T. l. 13690.

The plant intended was the clove-pink, -gilofre, or -gilliflower, Dianthus Caryophyllus, L.

SORB, L. *sorbus*, from *sorbeo*, drink down, in allusion to a beverage made from the fruit. See SERVICE-TREE. A name formerly given to Pyrus domestica, L. at present to Pyrus torminalis, L.

SORREL, Fr. *surelle*, a dim. derived from L. Germ. *suur*, sour, from the acidity of the leaves, Rumex Acetosa, L.

    ,,   SHEEP'S-,   Rumex Acetosella, L.

    ,,   WOOD-,   Oxalis Acetosella, L.

SOURINGS and SWEETINGS, crabs and sweet apples.

SOUTHERNWOOD, A.S. *suðernewude,* abridged from *superne wermod,* southern wormwood, as in Lib. Med. (O. Cockayne, Leechd. i. p. 51).

Artemisia Abrotanum, L.

SOWBANE, from being, as Parkinson tells us (Th. Bot. p. 749), " found certain to kill swine,"

Chenopodium rubrum, L.

SOWBREAD, G. *saubrodt,* L. *panis porcinus,* from its tuber being the food of wild swine,

Cyclamen europæum, L.

Sow THISTLE, in Pr. Pm. *thowthystil,* A.S. *pufepistel,* or *pupistel,* O.G. *du-tistel,* sprout thistle, from *pufe,* a sprout, an indication of the plant having been valued for its edible sprouts, which Evelyn tells us were eaten by Galen as a lettuce, and, as we learn from Matthioli (l. ii. c. 124), they were by the Tuscans, even in his day : " Soncho nostri utuntur hyeme in acetariis." It seems to have been called *sow-thistle,* through its name in the Ortus Sanitatis (c. cxlviii.) *suwe-distel,* or, in some editions, *saw-distel,* a corruption of its A.S. and older German name.

Sonchus oleraceus, L.

SOWD-WORT, the soda-plant, the plant from the ashes of which soda is obtained, Fr. *soude,* L. *solida,* soda being the solid residue left by boiling a lye of its ashes,

Salsola Kali, L.

SOWER, WOOD-, see SORREL.

SPARAGUS, in Evelyn's Acetaria, shortened from Lat. *Asparagus,* as *Emony* from *Anemony,* by the mistake of the initial vowel for the indefinite article, *a* or *an,* and still further corrupted to SPARROW-GRASS ; an example of the habit of the uneducated to explain an unknown word by a more familiar one ;   Asparagus officinalis, L.

SPARROW-TONGUE, from its small acute leaves, the knot-grass,   Polygonum aviculare, L.

Spart-grass, in the Northern counties (Brockett), "a dwarf rush common on moors and wastes," Sp. *esparto*. See Spurt-grass.   Spartina stricta, Sm.

Spatling-poppy, from A.S. *spatlian*, froth, from the spittle-like froth often seen upon it caused by the bite of an insect,   Silene inflata, L.

Spear-grass, in Shakspeare (Henry IV. 1st part, a. ii. sc. 4), and in Lupton's Notable Things, a plant used to tickle the nose and make it bleed, perhaps the common reed,   Phragmites communis, Trin.

Spear-mint or Spire-mint, from its spiry, not capitate inflorescence,   Mentha viridis, L.

Speedwell, from its corolla falling off and flying away as soon as it is gathered; "Speed-well!" being equivalent to "Farewell!" "Good-bye!" and a common form of valediction in old times. "Forget-me-not," a name that has since passed to a myosotis, appears to have first been given to this plant, and addressed to its fleeting flowers. See Forget-me-not.   Veronica Chamædrys, L.

Spelt, the same word in Du. G. Da. and Sw. the It. *spelda*, Sp. *espelta*, Fr. *espeautre*, from G. *spalten*, split. *Spelt* is explained in Levin's Manipulus by *eglumare*, to husk.   Triticum Spelta, L.

Sperage, Fr. *esperage*, from L. *asparagus*,
    Asparagus officinalis, L.

Spikenel, Spicknel, or Spignel, Sp. *espiga*, spike, and *eneldo*, from L. *anethum*, dill, a plant that was imported from Spain under that name; see Lyte, b. iii. c. 15;
    Meum athamanticum, L.

Spinach, It. *spinace*, derived, according to Diez and Scheler, through a presumed M.Lat. *spinaceus*, spiny, from L. *spina*, a thorn, in allusion to its sharp-pointed leaves, or, as others with more reason say, to its prickly fruit. If we assume a word for which we have no authority, *spinax* would bring us nearer to the It. *spinace* than *spinaceus*.

The analogy of other plant-names would suggest a M.Lat. *spinago*. But the word seems to have an entirely different origin. Fuchs tells us (Hist. Stirp. p. 668), that it is called in Arabic *Hispanach :* " Arabicæ factionis principes *Hispanach,* hoc est, Hispanicum olus nominant." Dodoens (b. v. i. 5) tells us also, " *Spinachiam* nostra ætas appellat, nonnulli *spinacheum* olus. Ab Arabibus et Serapione *Hispanac* dicitur." Brunfelsius (ed. 1531) says expressly at p. 16, " Quæ, vulgo *spinachia* hodie, Atriplex *Hispaniensis* dicta est quondam ; eo quod ab Hispania primum allata est ad alias exteras nationes." Tragus also calls it *Olus Hispanicum ;* Cotgrave *Herbe d'Espaigne ;* and the modern Greeks σπαναχιον. It is only in deference to the very high authority of Diez, that it has seemed necessary to quote these ancient authors. Talbot in Engl. Etym. takes the same view, and considers the name as meaning " Spanish." Spinacia oleracea, L.

SPINDLE-TREE, from its furnishing wood for spindles, A.S. *spindel,* which meant, not so much the implement used in spinning, as a pin or skewer, a purpose for which it is used to this day, and whence it has taken its other names of Gadrise, Prickwood, etc.

Evonymus europæus, L.

SPINKS, or BOG-SPINKS, Du. *pinkster-bloem,* from their blossoming at *pinksten* or Pentecost, Gr. πεντηκοστη (see PINK), the Lady's smock, Cardamine pratensis, L.

SPIRES, or SPIRE-REED, the pool-reed, A.S. *pol-spere,* in the Wycliffite version of Is. xix. 6, called *spier,* and in the Owl and Nightingale, l. 19, *spire ;*

"I-meind mid *spire* and grene segge."

Probably it meant a spear, A.S. *spior* or *spere,* and perhaps in the first place was so named from the Spanish reed, Arundo Donax, having been imported and used for missiles. In later times we find this word in the sense of a pointed inflorescence, as a " spyre of corne," Palsg. " I spyre as corne doth." ib. *Spire* is in different counties applied to several different plants, such as rushes and sedges. It usually means the common reed, Arundo Phragmites, L.

SPLEEN-WORT, from its supposed efficacy in diseases of the spleen, Gr. σπλην, a notion suggested, on the doctrine of signatures, by the lobular form of the leaf in the species to which the name was first given, the ceterach. See MILTWASTE. Asplenium, L.

SPOONWORT, G. *löffel-kraut*, from its leaf being shaped like an old-fashioned spoon; whence also its Latin name;
Cochlearia officinalis, L.

SPREUSIDANY, from L. *peucedanum*,
Peucedanum officinale, L.

SPRING-GRASS, see VERNAL-GRASS.

SPRUCE, from G. *sprossen*, a sprout, as the tree from the sprouts of which *sprossen-bier*, our spruce-beer, is made. Evelyn, from the expression he uses: "Those of Prussia, which we call Spruse," seems to have fancied that it meant "Prussian." Pinus Abies, L.

SPURGE, Fr. *espurge*, L. *expurgare*, from its medicinal effects, Euphorbia, L.

„ CAPER-, from its immature fruit being substituted for the real caper, Euphorbia Lathyris, L.

„ SUN-, from its flowers turning to the sun,
Euphorbia helioscopia, L.

SPURGE LAUREL, Daphne Laureola, L.

SPURGE OLIVE, Daphne Mezereon, L.

SPUR-WORT, It. *speronella*, from its verticils of leaves resembling the large spur-rowels formerly worn,
Sherardia arvensis, L.

SPURRY, a word from which Lyte says (b. i. ch. 38) that the Lat. *spergula* was formed. It seems more likely that *spergula* is contracted from *asparagula*, a presumed dim. of *asparagus*, a plant which it somewhat resembles, and *spurry* from *spergula*. The G. *spark*, and Fr. *espargoutte*, seem to

be the same word differently developed. Cotgrave gives a
Fr. *spurrie*. Spergula arvensis, L.

SPURT-GRASS, a rush of which the baskets were made,
that were called in A.S. *spyrtan*, and which seem, from one
of Ælfric's colloquies, to have been used for catching fish.
This word *spyrta* has probably been formed from L. *sporta*,
a basket made of *spartum*, the Sp. *esparto*, the grass so
much used for mats and baskets in the South, and related
to Gr. σπειραω, twist, wreathe.
Scirpus lacustris and maritimus, L.

SQUILL, L. *scilla* or *squilla*, Gr. σκιλλα. The same word,
the It. *squilla*, is now used to mean the small evening bell
sounded from the campanili in Italy for vespers service,
and this Diez would derive from O.H.G. *skilla*, G. *schelle*,
and the verb *skellan*, ring, and quotes a passage from the
Lex. Sal. " Si quis *schillam* de caballo furaverit," to show
its original use as a horse-bell. It seems far more probable
that the little bell should have been so called from its
resemblance to the bulb of an Italian plant, and its name
have been adopted by other nations with the Christian
religious rites, than that Italians should have first learnt a
name for such an old invention from the Germans. But
be the origin of *squilla* what it may, the flower was not
called so from any resemblance to a bell, as its synonym
" Harebell" might lead us to suppose, but is simply the
Gr. word σκιλλα. Scilla nutans, etc. Sm.

SQUINANCY, from its supposed efficacy in curing the
disease so called in old authors, viz. the quinsy, Fr. *esqui-
nancie*, M.L. *squinancia*, It. *schinanzia*, Gr. κυναγχη, from
κυων, dog, and ἀγχω, strangle, a dog-choking disease, one
in which the patient, from inflammation and swelling of
the fauces, is obliged to gasp with his mouth open like a
strangled dog, Asperula cynanchica, L.

SQUINANCY BERRIES, black currants, from their use in
sore throat, Ribes nigrum, L.

SQUIRREL TAIL, from the shape of the flower-spike,
Hordeum maritimum, With.

SQUITCH, or QUITCH, A.S, *cwice*, from *cwic*, vivacious,
the couch-grass, so called from its tenacity of life,
Triticum repens, and Agrostis stolonifera, L.

STAB-WORT, the wood-sorrel, so called, according to Parkinson (Th. Bot. p. 747), " because it is singular good in all wounds and stabbes into the body." By most authors it is spelt *stubwort*. Oxalis acetosella, L.

STAGGERWORT, usually understood to be so called from curing the staggers in horses, but to judge from its synonym *Seggrum*, and its being found in some works spelt *Staggwort*, more probably derived from its application to newly-castrated bulls called *Seggs* and *Staggs*,
Senecio Jacobœa, L.

STANDERWORT, or STANDERGRASS, Fl. *standelkruid*, G. *stendel-wurtz*, Sw. *standört*, names of which it is needless to unveil the meaning, but descriptive of a supposed effect of the " Foul standergrass," suggested by its double tubers, which, on the doctrine of signatures, indicate aphrodisiac virtues, Orchis mascula, L.

STANMARCH, O.E. *stane*, stone, and *march*, parsley, a translation of Gr. πετροσελινον, the Alexander,
Smyrnium Olusatrum, L.

STAR-FRUIT, from the radiated star-like growth of its seed-pods, Actinocarpus Damasonium, L.

STAR-GRASS, a grassy-looking aquatic plant with stellate leaves, Callitriche, L.

STAR HYACINTH, from its open stellate flowers,
Scilla verna, Hud.

STAR-JELLY, the nostoc, a jelly-like alga popularly supposed to be shed from the stars, Tremella Nostoc, L.

STAR THISTLE, from its spiny involucre, resembling the weapon called a morning star, Centaurea solstitialis, L.

STAR-WORT, from the shape of the flower,
Aster Tripolium, L.

STAR OF BETHLEHEM, from its white stellate flowers, like pictures of the star that indicated the birth of Jesus,
Ornithogalum umbellatum, L.

STAR OF THE EARTH, from its leaves spreading on the ground in star fashion,   Plantago Coronopus, L.

STAR OF JERUSALEM, It. *girasole*, turn-sun, its Italian name familiarized into *Jerusalem*, the salsify,
Tragopogon porrifolius, L.

STARCH-CORN, from starch being made of it,
Triticum Spelta, L.

STARCH-WORT, from its tubers yielding the finest starch for the large collars worn in Queen Elizabeth's reign,
Arum maculatum, L.

STARE, or STARR, Dan. *star*, or *stär-gräs*, Ic. *stör*, Sw. *starr*, words meaning "stiff grass," as in Douglas's Virgil, b. vi. l. 870:

"rispis harsk and *stare*,"

a name applied to various sedges and coarse sea-side grasses, more especially   Carex arenaria, L.
and   Ammophila arundinacea, Host.

STAVER-WORT, from being supposed to cure the stavers or staggers in horses,   Senecio Jacobæa, L.

STAVESACRE, a plant that was once in great use for destroying lice, but which with the gradual increase of cleanly habits is become scarce in our gardens, L. *staphisagria*, Gr. of Galen ἀσταφισαγρια, from ἀσταφις, raisin, and ἀγρια, wild, referring to the similarity of its leaf to that of the vine; unless Galen's plant was an entirely different one, for which ours has been mistaken;
Delphinium Staphisagria, L.

STAY-PLOUGH, the rest-harrow,   Ononis arvensis, L.

STICKADOVE, a name corrupted from the officinal Lat. *flos stoechados*, flower of the *stoechas*, a lavender so called from growing on the Hyeres, islands opposite Marseilles, and called by the ancients *Stoechades,* from standing in a row, Gr. στοιχας,   Lavandula Stoechas, L.

STINKHORN, from its shape and offensive odour,
                          Phallus impudicus, L.

STITCH-WORT, in a thirteenth century MS. in Mayer and Wright, p. 140, spelt *Stich-wurt,* and given as the translation of "Valeriane," a plant used to cure the sting, G. *stich,* of venemous reptiles; but in later works explained as curing the stitch in the side. See Gerarde, p. 140.
                          Stellaria Holostea, L.

STOCK-GILLIFLOWER, now shortened to STOCK, from *stock,* the trunk or woody stem of a tree or shrub, added to *Gilliflower* to distinguish it from plants of the Pink tribe, called, from their scent, *Clove-Gilliflowers,*
                          Matthiola incana, L.

STOCK-NUT, from its growing on a stick, G. *stock,* and not on a tree like the walnut,   Corylus Avellana, L.

STONE BASIL, a basil that grows among stones,
                          Calamintha Clinopodium, Benth.

STONE-BREAK, G. *steinbrech,* from L. *saxifraga,* so named from its supposed power of rending rocks, and thence employed to break stone in the bladder,   Saxifraga, L.

STONE-CROP, from *crop,* a top, a bunch of flowers, a cima, and *stone;* being a plant that grows on stone walls in dense tufts of yellow flowers;   Sedum acre, L.

STONE-FERN, from its growth on stone walls,
                          Ceterach officinarum, W.

STONE-HOT, or STONNORD, corruptions of *stone-wort,* and STONE-HORE, or STONOR, of *stone-orpine,* (see ORPINE,)
                          Sedum reflexum, L.

STONE-WORT, from calcareous deposits on its stalk,
                          Chara, L.

STONES, a translation of Gr. ὀρχις, a name given to several orchideous plants from their double tubers, and in old herbals used with the name of some animal prefixed, as, e.g. that of the dog, fox, goat, or hare,   Orchis, L.

STRANGLE-TARE, a tare that strangles,

Vicia lathyroides, L.

and also a plant that strangles a tare,

Cuscuta europæa, L.

STRAP-WORT, L. *corrigiola*, a little strap, dim. of L. *corrigia*, so called from its trailing habit,

Corrigiola littoralis, L.

STRAWBERRY, A.S. *streowberie*, either from its straw-like halms, or from their lying strown on the ground. Some have supposed that the name is derived from the custom in some parts of England to sell the wild fruit threaded on grass-straws. But it dates from a time earlier than any at which wild strawberries are likely to have been marketable.

Fragaria vesca, L.

STRAWBERRY CLOVER, from its round pink strawberry-like heads of seed, formed by the inflated calyx,

Trifolium fragiferum, L.

STRAWBERRY-TREE, from the shape and colour of its fruit,   Arbutus Unedo, L.

STUBWORT, from its growing about the stubs of hewn trees, the wood-sorrel,   Oxalis Acetosella, L.

STURTION, a corruption of L. *nasturtium*, a cress, a popular name of a plant which from the flavour of its leaves was by the old herbalists ranked with the cresses,

Tropæolum majus, L.

SUCCORY, Fr. *chicorée*, Gr. κιχωρη, the wild endive, too often replaced by fraudulent dealers with dandelion roots,

Cichorium Intybus, L.

SULPHUR-WORT, from its roots being, according to Gerarde (p. 1053), "full of a yellow sap, which quickly waxeth hard and dry, smelling not much unlike brimstone, called Sulphur,"   Peucedanum officinale, L.

SUNDEW, a name explained by Lyte in the following

description of the plant. "It is a herbe of a very strange nature and marvellous: for although that the Sonne do shine hoate and a long time thereon, yet you shall finde it alwayes moyst and bedewed, and the small heares [hairs] thereof alwayes full of little droppes of water: and the hoater the Sonne shineth upon this herbe, so much the moystier it is, and the more bedewed, and for that cause it was called *Ros Solis* in Latine, whiche is to say in Englishe, The dewe of the Sonne, or Sonnedewe." Nevertheless, the Germ. name, *sindau*, leads us to suspect that the proper meaning of the word was "ever-dewy," from A.S. O.S. and Fris. *sin*, ever, rather than from *sun*. The Latin name, *Ros solis*, is modern, and, as the plant is seldom met with in the South of Europe, is probably a mistranslation of the German or English one.   Drosera, L.

SUNFLOWER, from its "resembling the radiant beams of the sun," as Gerarde says; or, as another old herbalist expresses it in Latin, "idea sua exprimens solis corpus, quale a pictoribus pingitur;" and not, as some of our popular poets have supposed, from its flowers turning to face the sun, which they never do; a delusion that Thomson expresses in the lines:

> "But one, the lofty follower of the sun,
> Sad when he sets, shuts up her yellow leaves,
> Drooping all night, and, when he warm returns,
> Points her enamour'd bosom to his ray."
>
> Summer, l. 216.
>
> Helianthus annuus, L.

also in some herbals, from its only opening in the sunshine, the rock rose,   Helianthemum vulgare, Gärt.
in our older poets, the marigold, as in Heywood's Marriage Triumphe:

> "The yellow marigold, the sunne's own flower."

"It was so named," says Hyll, "for that after the rising of the sun unto noon, this flower openeth larger and larger;

but after the noontime unto the setting of the sun the flower closeth more and more, so that after the setting thereof it is wholly shut up." Hyll, Art of Gard. c. xxx.

"The Marigold observes the sun,
More than my subjects me have done."
K. Charles I.

This is also the flower that in Anglo-Saxon is called *solsæce*, Fr. *souci*, from O.F. *soucicle*, L. *solsequium*, sun-following.
Calendula officinalis, L.

SUN-SPURGE, from its flowers turning to face the sun,
Euphorbia helioscopia, L.

SWALLOW-PEAR, a wild pear that is not a "choke-pear," a kind that may be eaten, Pyrus torminalis, L.

SWALLOW-WORT, Gr. χελιδονιον, of χελιδων, a swallow, because, according to Pliny (b. xxv. ch. 8), it blossoms at the season of the swallow's arrival, and withers at her departure, a name, that, for the same reason, has been given to several other plants, as the Ranunculus Ficaria, Fumaria bulbosa, Caltha palustris, and Saxifraga granulata, L. ; but, according to Aristotle and Dioscorides, because swallows restore the eyesight of their young ones with it, even if their eyes be put out. It is to be recollected, that, however absurd some of these superstitions, they may nevertheless be the real source of the name of a plant. Chelidonium majus, L.

SWEDE, a turnip so called from having been introduced from Sweden, Brassica campestris, L. var. rutabaga.

SWEET ALISON, a plant with the smell of honey, a species of the genus *Alyssum*, of which *Alison* is a corruption, and not the name of a pretty lady,
Alyssum maritimum, L.

SWEET BAY, from the odour of its leaves, to distinguish it from other evergreen shrubs, such as the strawberry tree and cherry laurel, that were once reckoned among the bays, Laurus nobilis, L.

Sweet-briar, a wild rose whose leaves are sweet-scented,
Rosa rubiginosa, L.

Sweet Chervil, or Sweet Cicely, from its agreeable scent, Gr. σεσελι, Myrrhis odorata, Scop.

Sweet Flag, to distinguish it from the unscented flag, or iris, Acorus Calamus, L.

Sweet Gale, from its scent, Myrica Gale, L.

Sweet John, probably a fanciful name given to certain varieties of pink to distinguish them from those called Sweet Williams. They seem to have been the narrow-leaved kinds. Dianthus barbatus, L.

Sweet-pea, a scented pea, Lathyrus odoratus, L.

Sweet-sedge, or -seg, a plant which, having sword-blade leaves, was comprised under the general name of *Segs* and *Sedges*, and fraudulently sold in shops for the sweet cane or calamus aromaticus,
Acorus Calamus, L.

Sweet William, from Fr. *oeillet*, L. *ocellus*, a little eye, corrupted to *Willy*, and thence to *William*, in reference, perhaps, to a popular ballad; a name assigned by W. Bulleyn (fol. 48), to the wallflower, but by later herbalists to a species of pink. See William.
Dianthus barbatus, L.

Sweet Willow, from its having the habit of the dwarf willows, and sweet-scented foliage, the sweet gale,
Myrica Gale, L.

Sweeting, a sweet apple, as contrasted with the crab,
Pyrus Malus, L.

Sweth, L. Germ. of Turner's time *Suitlauch*, perhaps a misprint of *snitlauch*, or, as it is given in a German edition of Macer (ed. 1590), *snithlauch*, properly *schnittlauch*, a garlick to be cropped and grow again, chives,
Allium Schœnoprasum, L.

Swine's-bane, see Sowbane.

Swine's-cress, a cress only good for swine,
Senebiera Coronopus, Poir.

Swine's-grass, *Swynel grass* of the Grete Herball, Da. *swinegræs*, the knotgrass. Johnston in East. Bord. observes that "Swine are said to be very fond of it," a statement confirmed by writers on agriculture.
Polygonum aviculare, L.

Swine's-snout, L. *rostrum porcinum*, from the form of the receptacle, the dandelion, Taraxacum Dens leonis, L.

Swine Succory, a translation of its Greek name from ύος, pig's, and σερις, succory, Hyoseris minima, L.

Sword Flag, from its banner-like flower, and sword-shaped leaf, Iris Pseudacorus, L.

Sycamine, in old authors the woodbine.

Sycamore, Gr. συκομορος, properly the name of the wild fig, but by a mistake of Ruellius, according to J. Bauhin (Hist. Plant. p. 168), transferred to the great maple. This mistake arose, perhaps, from this tree, the great maple, being, on account of the density of its foliage, used in the sacred dramas of the Middle Ages to represent the fig-tree into which Zaccheus climbed, and that in which the Virgin Mary, on her journey into Egypt, had hidden herself and the infant Jesus, to avoid the fury of Herod ; a legend quoted by Stapel on Theophrastus (p. 290, a), and by Thevenot in his Voyage de Levant (part i. p. 265) : "At Matharee is a large sycamore or Pharaoh's fig, very old, but which bears fruit every year. They say that upon the Virgin passing that way with her son Jesus, and being pursued by the people, this fig-tree opened to receive her, and closed her in, until the people had passed by, and then opened again." The tree is still shown to travellers. (See Cowper's Apocryphal Gospels, p. 191, note.) The great maple was naturally chosen to represent it, from its making, as W. Gilpin expresses it, "an impenetrable shade." Acer Pseudoplatanus, L.

The *Sicamour* of Chaucer in his Flower and Leaf (l. 54) was some twining shrub, probably the honeysuckle:

> "The hegge also that yede in compas,
> And closed in all the greene herbere,
> With *sicamour* was set and eglatere,
> Wrethen in fere."

and l. 66,

> "The hegge as thicke as any castle wall,
> That who that list without to stond or go,
> Though he would all day prien to and fro,
> He should not see if there were any wight
> Within or no."

Lonicera caprifolium, L.

SYNDAW, G. *sindau*, constant dew, in Fuchs and the Ortus Sanitatis *sinnau*, a name at present confined to the sun-dew, but by Wm. Turner (b. iii. p. 24) given to the Lady's mantle, both these plants having formerly been comprehended by Cordus and others under that of Drosera,

Alchemilla vulgaris, L.

SYRINGA or SYRING, a name commonly given to a shrub whose stems are used in Turkey for making pipe-sticks, from L. *Syrinx*, a nymph who was changed into a reed,

Philadelphus coronaria, L.

by Evelyn to the lilac, that for the same reason was called Pipe tree, 　　　　　　　　　　Syringa vulgaris, L.

TANG, O.N. *þang*, Da. *tang*, Fris. *mar-tag*, a word that corresponds to Da. *tag*, A.S. *þæce*, thatch, from sea-weed having formerly been used to cover houses, instead of straw. The word has been adopted from one of the northern languages, and refers to a time earlier than the cultivation of cereal grains in high latitudes. Fucus nodosus, L.

TANGLE, seemingly an attempted explanation of *Tang*, as if it meant *entangling*, 　　　Laminaria digitata, Ag.

TANSY, Fr. *athanasie*, now contracted to *tanacée* and *tanaisie*, M.Lat. *athanasia*, the name under which it was sold in the shops in Lyte's time, Gr. ἀθανασια, immortality, referring to a passage in Lucian's Dialogues of the Gods (no. iv.), where Jupiter, speaking of Ganymede, says to Mercury, ἀπαγε αὐτον, ὦ Ἑρμη, και πιοντα της ἀθανασιας ἀγε οἰνοχοησαντα ἡμιν. "Take him away, and when he has drunk of immortality, bring him back as cupbearer to us." The ἀθανασια here has been misunderstood, like ἀμβροσια in other passages, for some special plant. Dodoens says (l. i. 2, 16), that it was called so, " quod non cito.flos inarescat," which is scarcely true. Hyac. Ambrosinus, in his Phytologia, p. 82, says : "Athanasia ita vocata quia ejus succus vel oleum extractum cadavera a putredine conservat." *Tanacetum*, its systematic name, is properly a bed of tansy, and is a word of modern origin.

Tanacetum vulgare, L.

,, Goose-, or Wild-, from its tansy-like leaves,

Potentilla anserina, L.

Tare, an obscure word for which many derivations have been proposed. In old works it is usually combined with *fytche*, as the *tare*-fytche. The word *tire-lupin* in Rabelais' preface seems to explain it as derived from Fr. *tirer*, drag, and to mean a vetch that pulls other plants towards it.

Lathyrus, Ervum, Vicia.

Tarragon, a corruption of its Lat. specific name, meaning " a little dragon,"   Artemisia Dracunculus, L.

Tassel-grass, a grass-like plant with bunches of delicate leaves like tassels,   Ruppia maritima, L.

Teasel, A.S. *tæsel*, from *tæsan*, tease, applied metaphorically to scratching cloth,   Dipsacus fullonum, L.

Tench-weed, either from its growing in ponds where tench have broken up the puddling by burrowing in it ; or, as Forby says, " from its having a slime or mucilage about it that is supposed to be very agreeable to that fish ;"

Potamogeton natans, L.

Tent-wort, the wall rue, a fern so called from its having

been used as a specific or sovereign remedy in the cure of rickets, a disease once known as the *Taint*. Threlkeld, under Adiantum album, says : "It is one of the capillary plants, and a specific against the Rickets. For this reason our ancestors gave it the name of *Tent-wort*."

Asplenium Ruta muraria, L.

TETTER-BERRY, from its curing a cutaneous disease called *tetters*, Bryonia dioica, L.

TETTER-WORT, from its curing tetters,

Chelidonium majus, L.

TEYL-, TEIL-, TIL-, or TILLET-TREE, the lime, Fr. *tille*, formerly spelt *teille*, a word now confined to the inner bark or bast of the tree, and replaced with the dim. *tilleul*, from M. Lat. *tilliolus*, dim. of *tilia*, Tilia europæa, L.

THALE-CRESS, from a Dr. *Thalius*, who published a catalogue of the plants of the Hartz mountains,

Arabis Thaliana, L.

THAPE, see FEABE.

THEVE-THORN, O.E. of Pr. Pm. *thethorn*, A.S. *þefe-*, *þife-*, or *þyfe-þorn*, a word that occurs in Wycliffe's Bible, in the fable of Jotham (Judg. ix. 14, 15), as a translation of the L. *rhamnus* of the Vulgate, Heb. *atad*, the name that Dioscorides, as cited by Bochart (i. 752), says that the Carthaginians also called a large species of *rhamnus*. It is unknown what bramble Wycliffe meant. T. Wright, in his Manners of the Middle Ages, p. 296, takes it for the *Thape* or gooseberry. The context requires a barren or worthless brier, and the monks who commented upon Mesues took it to be the *dewberry :* " Monachi qui in Mesuem commentarios edidere, *Rhamnum* existimaverunt rubum quendam, qui humi repens, incultisque proveniens, mora cæruleo potius quam nigro colore profert." Matthioli (l. i. c. 102). They probably followed an ecclesiastical tradition, in fixing upon this particular bramble, as representing *rhamnus*. The word *theve* seems to be related to such as imply lowliness and subservience, Go. *þivan*, to subject, *þivi*, a female slave, etc. (See Diefenbach, Lex. comp. ii. 708.) In this view of it *Theve-thorn* or *Theue-thorn*, as we find it printed, will be the parent of *Dew-berry* rather than of *Thape*, and I have no hesitation in referring it to that species.   Rubus cæsius, L.

THISTLE, A.S. *þistel*, from *þydan*, stab, and the same word essentially in all the kindred languages.

,,   BLESSED-, from its use against venom,
                   Carduus benedictus, L.

,,   CARLINE-, from its curing Charlemagne's army of a pestilence,   Carlina vulgaris, L.

,,   CORN-, or WAY-, from its growing in fields,
                   Serratula arvensis, L.

,,   COTTON-, from its cottony white stems and leaves,   Onopordon Acanthium, L.

,,   FULLER'S-, from its use in dressing cloth,
                   Dipsacus Fullonum, L.

,,   GENTLE-, from its comparatively soft, unarmed, and inoffensive character,   Carduus Anglicus, L.

,,   HOLY-,   Carduus benedictus, L.

,,   MELANCHOLY-, from its use in hypochondria,
                   Carduus heterophyllus, L.

,,   MUSK-, from the scent of its flowers,
                   Carduus nutans, L.

,,   OUR LADY'S-, from being dyed with her milk,
                   Carduus Marianus, L.

,,   ST. BARNABY'S-, from its season of flowering,
                   Centaurea solstitialis, L.

,,   SCOTCH-, as being the badge of Scotland,
           Onopordon Acanthium, and Carduus nutans, L.

,,   SOW-, a mistake of its A.S. name *þuþistel*,
                   Sonchus oleraceus, L.

,,   STAR-, from its star-shaped involucre,
                   Centaurea Calcitrapa, L.

THORN, A.S. *þorn*, Go. *þaurnus*, and, like *thistle*, the same word in all the kindred languages, and used with it alliteratively, extending to the Slavonian and Celtic dialects also, related perhaps to Gr. τορεω, bore, L. *terebro*, Skr. *tri*, pass through, Boh. *trn*, Pol. *tarn*, Wel. *draen*, etc., a word of unknown derivation.

„	BLACK-, the sloe,	Prunus spinosa, L.
„	BUCK-,	Rhamnus catharticus, L.
„	HAW-,	Cratægus oxyacantha, L.
„	SALLOW-, or WILLOW-,	
		Hippophae rhamnoides, L.
„	WAY-,	Rhamnus catharticus, L.
„	WHITE-,	Cratægus oxyacantha, L.

THORN-APPLE, a plant with a thorny fruit,

Datura Stramonium, L.

THORN-BROOM, the furze, Ulex europæus, L.

THOROW-WAX, or THROW-WAX, a name given to the plant by Turner, because, as he says, "the stalke waxeth throw the leaves," Bupleurum rotundifolium, L.

THREE-FACES-UNDER-A-HOOD, the pansy,

Viola tricolor, L.

THRIFT, the passive participle of *threave* or *thrive*, press close together, and meaning the "clustered" pink, so called from its growing in dense tufts, Armeria vulgaris, W.

THROAT-WORT, G. *halswurz*, the Canterbury bell, from being supposed, from its throat-like corolla, to be a cure for sore throats, Campanula Trachelium, L.

THRUM-WORT, from *thrum*, a warp-end of a weaver's web, as in the Teesdale proverb, "He's nae good weaver that leaves lang *thrums*," a word used by Lyte in describing the reed-mace, the head of which he says (b. iv. c. 53), "seemeth to be nothing els but a throm of gray wool, or flockes, thicke set and thronge togither." The plant has its name from its long tassel-like panicles of red flowers.

Amaranthus caudatus, L.

THYME, Gr. θυμος, from θυω, fumigate, and identical with L. *fumus*, Skr. *dhuma*, smoke, from *dhu*, agitate, the name of some odoriferous plant or shrub used in sacrifices, at present appropriated to a genus of labiate plants, and more particularly to the hill-thyme,

Thymus Serpyllum, L.

TILLET, in Ph. Holland's Pliny, the till or lime tree.

TILLS, abbreviated from *lentils*. "The country people," says Parkinson (Th. Bot. p. 1068), "sow it in the fields for their cattle's food, and call it *Tills*, leaving out the *Lent*, as thinking that word agreeth not with the matter. *Ita sus Minervam*." Ervum Lens, L.

TIMOTHY-GRASS, from its having been brought from New York by a Mr. Timothy Hanson, and introduced by him into Carolina, and under that name first recommended to the attention of English agriculturists,

Phleum pratense, L.

TINE-TARE, a tare that *tines*, or encloses and imprisons other plants, Vicia hirsuta, Koch.
and in Linn. Soc. Journ. vol. v.   Lathyrus tuberosus, L.

TITHYMALL, a name of the spurge tribe in old writers, L. *tithymalus*, Euphorbia, L.

TOAD-FLAX, from its narrow, linear, flax-like leaves, and its having been described by Dodoens, as "Herba assimilis cum *Bubonio* facultatis," and *Bubonio* having been mistaken for *bufonio*, from *bufo*, a toad; as it is in the Ortus Sanitatis, ch. 431, where the author, speaking of the *Bubonium*, says, "dieses kraut wird von etlichen genennet *bubonium*, das ist *kroten*-kraut; wann *bubo* heisset ein *krot*. Inde *bubonium*." Linaria vulgaris, L.

„ BASTARD-, from its leaves resembling those of the preceding plant, Thesium linophyllum, L.

TOAD-PIPE, from its straight fistulous stalks, and growth in damp places where toads haunt, and croak, and pipe to one another, Equisetum limosum, L.

TOAD-STOOL, any of the unwholesome fungi, from a popular belief that toads sit on them. So Spenser, in Sheph. Cal. Dec. 1. 69 :

> " The griesly todestool grown there mought I see,
> And loathed paddocks lording on the same."
>
> Boletus, Agaricus, etc.

TOBACCO, a name of the plant that was adopted by the Spaniards from the Indians of Cuba, and properly the name of the pipe in which it was smoked, the weed itself having been called *Cohiba*,　　　Nicotiana Tabacum, L.

TOLMENEER, TOLMEINER, or COLMENIER, a name given by the herbalists to a variety of the Sweet William. It is spelt by Lyte (b. ii. ch. 7), in three syllables *Tol-me-neer*, as though it meant " Toll or entice me near," and *Colme-nier* might in the same way be explained as " Cull me near." It is most likely a corruption of *D'Almagne*, or *D'Allemagne*, as being a pink of Germany.

Dianthus barbatus, L.

TOMATO, its American-Indian name, Sp. *tomate*,

Solanum Lycopersicum, L.

TOOTH-CRESS, or TOOTH-VIOLET, from the tooth-like scales of the root,　　　Dentaria bulbifera, L.

TOOTHWORT, from the tooth-like scales of the root-stock, and base of the stem,　　　Lathræa Squamaria, L.

TORCH, G. *dartsch*, the mullein, called so, because, according to Parkinson (Th. Bot. p. 62), and W. Coles (ch. 112), " the elder age used the stalks dipped in suet to burn, whether at funerals or otherwise." See HIGTAPER.

Verbascum Thapsus, L.

TORMENTIL, Off. L. *tormentilla*, from the use of the root in cases of dysentery, L. *tormina*, or, as Lobel says, from its relieving the pain of toothache,

Potentilla Tormentilla, Sibt.

TOUCH-ME-NOT, from the sudden bursting of its seed-pods, upon being touched ; a phrase that was familiar from the "noli-me-tangere" pictures in Roman Catholic countries;   Impatiens Noli me tangere, L.

TOUCH-WOOD, a fungus imported from Germany, and apparently called for that reason *Dutch-wood*,

Boletus igniarius, L.

TOWER-CRESS, from its having been found growing upon the tower of Magdalen College, Oxford. Its Lat. specific name, *turrita*, expresses a pyramidal habit of growth, and seems to have been given to it as a translation of its trivial English name, in mistake of its intended meaning.

Arabis Turrita, L.

TOWER-MUSTARD, from the tapering growth of the inflorescence something in the form of a Dutch spire, " om de spits torrewijse oft naeldewijze ghewas van de steelkens," says Lobel (Kruydtboek, p. 262),   Turritis glabra, L.

TOWN-CRESS, A.S. *tun-cærse*, a cress grown in a *tun*, or enclosed ground, as contrasted with the water- and other wild cresses. *Tun* seems usually to have meant, as in Iceland at the present day, a close or pasture in connexion with farm buildings. So in Wycliffe's N. Test. (Luke xiv. 18): " I have bought a *toune*." It is still retained in this sense in that very interesting and only half-explored magazine of antiquities, the nomenclature of our fields; and met with occasionally as *Tun-mead*. It is also the Du. *tuin*, a garden.   Lepidium sativum, L.

TOY-WORT, from the little imitations of purses that it bears,   Capsella Bursa, DC.

TRAVELLER'S JOY, from the shade and shelter that it affords by the bowers it forms in roadside hedges,

Clematis Vitalba, L.

TREACLE-MUSTARD, or -WORMSEED, from its being used among 72 other ingredients, in making " Venice treacle," a famous vermifuge and antidote to all animal poisons, and one that was in great vogue during the Middle Ages; Du. *triakel*, Off. L. *theriaca*, from. Gr. θηριον, a small animal.

It was so named from the vipers that were added to the Mithridate by Andromachus, physician to the Emperor Nero.

"Andromachus a voulu changer le nom de *Mithridate* en celuy de *Theriaque*, à cause des vipères, auxquelles il a attribué le nom de θηριον, et lesquelles il a ajouté pour la base principale de cette composition."—Chares, l'histoire des animaux, etc., qui entrent dans la Theriaque. Paris, 1668. See also Heberden's Antitheriaca.

This remedy, *triacle*, as it was called, was the source of many popular tales of sorcerers eating poison, and was retained in the London Pharmacopeia till about 100 years ago. It was to cure " all those that were bitten or stung of venomous beastes, or had drunk poison, or were infected with the pestilence." The name has extended into Persia, where opium is called *Teryak*. Vambéry (Central Asia, p. 428). See Pr. Pm., v. *Treacle*. Erysimum cheiranthoides, L.

TREACLE, POOR-MAN'S-,   Allium sativum, L.

TREE, A.S. *treow*, a word in which we find very much mixed up in different languages the meaning of a living tree, timber, and an oak-tree especially : Skr. *druma, druta, dru*, tree, *dâru*, wood ; Zend. *dru ;* Gr. δορυ, both a spearshaft and a tree, δρυς, an oak and a tree generally ; Slav. *drevo* and *dervo*, both wood and tree ; Alban. *dru*, wood ; Wel. *dar, derw*, oak ; Gael. *darach ;* Go. *triu*, A.S. *treow*, Da. *trä*, both wood and tree. These words seem to be related to Skr. *dhruva*, Zend. *drva*, firm, O.H.G. *triu*, true. It is certainly remarkable that at the early period of the first formation in Asia of our common mother-language, and long before the invention of steel, so hard a wood as that of the oak should have been so commonly used, as to have become synonymous with timber generally. Yet the most ancient boats that have been discovered are of this tree, hollowed out with the aid of fire, and tools of stone.

TREFOIL, L. *tria foliola*, three-leaflets, a name given more particularly to the clovers, Trifolium, L.

TREMBLING POPLAR, the aspen, from its quivering leaves.
Populus tremula, L.

TRIFFOLY, of Shakspeare's Cephalus and Procris, TRIFOLIE of others, a trefoil, so called from its three leaflets.

TRINITY, see HERB TRINITY.

TRIP MADAM, Fr. *trippe madame*, corrupted from *triacque madame*, a plant used as a *treacle* or vermifuge,
Sedum reflexum, L.

TROLL-FLOWER, the globe-flower, from Sw. *troll*, Da. *trold*, Fris. *trol*, a malignant supernatural being, a name corresponding to Scotch *Witches Gowan*, and given to this plant on account of its acrid poisonous qualities ;
Trollius europæus, L.

TRUELOVE, incorrectly so spelt for TRULOVE, a plant called so from its four leaves being set together in the form of a trulove-, or engaged lovers' knot, such as is seen in coats of arms where the wife's is quartered with her husband's ; from Da. *trolovet*, betrothed, of *tro*, faith, and *love*, promise, O.N. *trolofad*, and not from faithfulness in *love*, with which it has no etymological connexion ;
Paris quadrifolia, L.

TRUFFLE, in Parkinson and Dale TRUBBES, It. *tartuffola*, dim. of *tartufo*, from L. *terræ tuber*,
Tuber cibarium, Sibt.

TULIP, in old works TULIPAN, a word that in Turkish means "turban," Pers. *dulbend*, from its rich and varied colours, and its shape resembling that of an inverted cap,
Tulipa Gesneriana, L.

TUNHOOF, from A.S. *tun*, a court or garden, Du. *tuin*, and *hufe*, a crown, a translation of Gr. στεφανωμα γης, and L. *terræ corona*, the ground ivy,
Nepeta Glechoma, Benth.

TURK'S CAP, in Parkinson's Paradisus, the tulip, a translation of its Oriental name. See TULIP.

TURNIP, L. *terræ napus*, Brassica Rapa, L.

TURNSOLE or TORNSOLE, a name given in some old works to the wartwort, from its being supposed to turn its flowers towards the sun, Fr. *tournesol*, Gr. ἡλιοτροπιον,

Euphorbia helioscopia, L.

TUSSACK-GRASS, from its growing in thick tufts or tussocks, Aira cæspitosa, L.

TUTSAN, in old works TUTSAYNE, Fr. *toute-saine*, all-wholesome, a word that Gerarde (p. 435), Nemnich, and others derive from its "healing all." Thus Lobel tells us (Kruydtb. pt. i. p. 768), that " it is called by the common people in France *Toute saine*, because, like the Panacea, it cures all sicknesses and diseases." This idea has probably been suggested by its M. Lat. name *Androsæmum*, Gr. of Dioscorides (iii. 172) ἀνδροσαιμον, of ἀνδρος, man's, and αἱμα, blood, a name given to it in reference either to the claret colour of the juice of its ripe capsule ; or the blood-stain left on the fingers after rubbing the flower, as Fuchs explains it; or more probably to an unguent made from this and a closely-allied species, of which Gerarde says (p. 433) : "The leves, floures, and seeds stamped, and put into a glasse with oile olive, and set in the sunne for certain weekes, doth make an oile of the colour of blood, which is a most pretious remedy for deep wounds, and those that are thorow the body."

Hypericum Androsæmum, L.

TWAY-BLADE, from its two root-leaves,

Listera ovata, L.

TWICE-WRITHEN, L. *bistorta*, Polygonum Bistorta, L.

TWIG-RUSH, from its tough, twiggy, branching growth,

Cladium Mariscus, L.

TWITCH, see QUITCH-GRASS.

TWOPENCE, see HERB TWOPENCE.

TWOPENNY-GRASS, so called by Turner from its pairs of round leaves " standyng together of ech syde of the stalke lyke pence," Lysimachia Nummularia, L.

UNSHOE-THE-HORSE, It. *sferra-cavallo*, from its horse-shoe-shaped legumes being, upon the doctrine of signatures, supposed to have that power. " Wann die Pferde auf der Weide gehen, und sie auf diess kraut oft treten, fallen ihnen die Hufeisen bisweilen ab." Tabernæm. ii. p. 230.

Hippocrepis comosa, L.

„ „ also the moonwort, of which Du Bartas says, p. 79, ed. 1611 :

" Horses that feeding on the grassie hills,
Tread upon *Moonwort* with their hollow heels,
Though lately shod, at night go barefoot home."

Botrychium Lunaria, Sw.

UPSTART, the meadow saffron, from its flowers starting up suddenly from the ground, without putting out leaves first, Colchicum autumnale, L.

VALERIAN, L. *valeriana*, a name of uncertain origin,
Valeriana officinalis, L.

VELVET-DOCK, the mullein, from its large soft leaves,
Verbascum Thapsus, L.

VELVET-FLOWER, the love-lies-bleeding, from its crimson velvety tassels, Amaranthus caudatus, L.

VELVET-LEAF, Lavatera arborea, L.

VENUS'-BASON, L. *Veneris labrum*, Gr. Ἀφροδιτης λουτρον, the teasel, so named after the goddess of beauty, from the hollows formed by the united bases of the leaves being usually filled with water, that was used, says Ray, " ad verrucas abigendas," to remove warts and freckles,
Dipsacus sylvestris, L.

VENUS'-COMB, from the slender tapering beaks of the seed-vessels being set together like the teeth of a comb,
Scandix Pecten, L.

16

VENUS'-HAIR, the maidenhair fern, Adiantum, L.

VENUS'-LOOKING-GLASS, from the resemblance of its flowers set upon their cylindrical ovary to an ancient round mirror at the end of a straight handle. The name is given by Spenser (F. Q. iii. i. 8) to a magic mirror, in which a lady might see her destined husband.

Campanula hybrida, L.

VERNAL-GRASS, from its early flowering,

Anthoxanthum odoratum, L.

VERVAIN, Fr. *verveine*, L. *verbena*,

Verbena officinalis, L.

VETCH, or FETCH, or FITCH, It. *veccia*, L. *vicia*, related to *vincire*, bind, as the G. *wicke* to *wickeln*, from its twining habit. Used absolutely it means the common vetch,

		Vicia sativa, L.
,,	BITTER-,	Vicia Orobus, DC.
,,	GRASS-,	Lathyrus Nissolia, L.
,,	HORSE-SHOE-,	Hippocrepis comosa, L.
,,	KIDNEY-,	Anthyllis vulneraria, L.
,,	MILK-,	Astragalus glycyphyllos, L.
,,	TARE-,	Vicia hirsuta, Koch.

VETCHLING, a spurious vetch, Lathyrus pratensis, L.

VINE, Fr. *vigne*, L. *vinea*, adj. of *vinum*, wine, as being the wine-shrub, G. *wein-rebe*, Gr. οἶνος. Turner and some other old writers spell it *vynde*, from confusion with A.S. *winde*. The first syllable of *vinum* and *vitis* is probably a radical *wi* or *vi*, whence L. *viere*, twist, *with*, *withy*, etc., and given to the vine in reference to its twining habit.

Vitis vinifera, L.

VIOLET, It. *violetta*, dim. of L. *viola*, which itself is a dim. of ἴον; in botanical nomenclature now confined to the genus to which the pansy belongs, but by ancient writers extended to many other very different plants, especially scented ones. Indeed Laurenberg in Appar. Plant. says (p. 77) : "Videntur mihi antiqui suaveolentes quosque flores generatim *Violas* appellasse, cujuscunque etiam forent generis." Even to the present day we retain it in the popular names of several plants of very different orders. Used absolutely, it means the genus   Viola, L.

VIOLET, CORN-,   Campanula hybrida, L.
   ,, DAMASK-, or DAME'S-,
                          Hesperis matronalis, L.
   ,, DOG-,   Viola canina, L.
   ,, MARCH-,   Viola odorata, L.
   ,, TOOTH-,   Dentaria bulbifera, L.
   ,, WATER-,   Hottonia palustris, L.

VIPER'S BUGLOSS, a bugloss which, from its seed being like the head of that reptile, was supposed, on the doctrine of signatures, to cure its bite. Thus Matthioli (l. iv. c. 69): "In Echio, herba contra viperarum morsus celeberrima, natura semen viperinis capitibus simile procreavit."
                                   Echium vulgare, L.

VIPER-GRASS, L. *viperaria*, because, according to Monardus, a physician of Seville quoted in Parkinson's Th. Bot. (p. 410), " a Moore, a bondslave, did helpe those that were bitten of that venomous beast, the viper, which they of Catalonia called *Escuerso*, with the juice of this herbe, which both took away the poison, and healed the bitten place very quickly, when Treakle [Theriaca] and other things would do no good." Its Italian and officinal Latin name, *scorzonera*, is derived from It. *scorzone*, a venemous serpent, popularized into a word that would seem to mean " black rind," *scorza nera*.   Scorzonera edulis, Mn.

VIRGIN'S BOWER, a shrub so named by Gerarde, as fitting to be a bower for maidens, and with allusion, perhaps, to Queen Elizabeth, but not, as we might be tempted to imagine, to the Virgin Mary in a riposo, or resting scene on the way to Egypt, the frequent subject of pictures,   Clematis Vitalba, L.

WAGWANT, a West-country term supposed to mean *wag-wanton*, Fr. in Clusius (p. ccxviii), *amourettes tremblantes*, the quaking-grass, so called from its quivering spikelets,

Briza media, L.

WAKE-PINTLE, a name given in Florio and Torriano's Dictionary as the translation of Ital. *Aro*, and apparently identical in its meaning with *Wake-robin*, and *Cuckoo-pint*, the *wake* being, like *cuckoo*, a modern form of A.S. *cwic*, *cucu*, and Low. Germ. *quek*, alive, and the noun allusive to its supposed aphrodisiac powers. Arum maculatum, L.

WAKE ROBIN, from Fr. *robinet*, a word of the same meaning as *pintle*. See CUCKOO-PINT and WAKE-PINTLE.

Arum maculatum, L.

WALE-WORT, or WALL-WORT, the dwarf elder, A.S. *wealwyrt*, from A.S. *wal*, slaughter, or *wealh*, foreign, and corresponding to the other names of the plant, *Danesblood* and *Danewort*, which Aubrey tells us were given to it from its growing at a village called Slaughterford in Wiltshire, where it is supposed that an army of Danes was destroyed. In German *walwurz* means the comfrey. Ort. San. (c. xcv.). Brunsch. (b. ii. c. xx.).   Sambucus Ebulus, L.

WALNUT, or WELSH NUT, A.S. *wealh-hnut*, from *wealh*, foreign, G. *wälsch*, O.H.G. *walah*, Fr. *gauge*, an adjective used more particularly of Italy, from whence the tree was introduced into Northern Europe,   Juglans regia, L.

WALL BARLEY, a barley that grows about walls, and which seems to have been taken for the species called by Pliny (l. xxii. c. 25) *lolium murinum*, mouse-darnel, the *murinum* of which was confused with *murale*, and understood as *wall*-darnel. Thus Tragus explains the name, as given to the plant, " weil es von sich selbst auf den *Mawren* wächst;" and Turner tells us (part ii. p. 17) : " It is called of the Latines *Hordeum murinum*, that is, Wall-barley." See MOUSE-BARLEY.   Hordeum murinum, L.

WALL CRESS,   Arabis, L.

WALL FERN, Polypodium vulgare, L.

WALL-FLOWER, a plant introduced from Spain as a *Wall Stock-gillofer*, which became successively *Wall gilliflower*, and *Wall-flower*. The *Gillofer* was the French *giroflier*, and under *Stock-gillofer* was comprehended the *Stock*, Matthiola incana, as well as the wall-flower.

Cheiranthus Cheiri, L.

WALL PENNYWORT, from its round leaves,

Cotyledon Umbilicus, L.

WALL PEPPER, from its biting taste,   Sedum acre, L.

WALL ROCKET, from its rocket-like leaves,

Brassica tenuifolia, Boiss.

WALL RUE, a fern so called from its rue-like leaves,

Asplenium Ruta muraria, L.

WARE, A.S. *war*, sea-weed generally, Algæ, L.

WARENCE, Fr. *garance*, M.L. *varantia*, or *verantia*, the madder, from *vera*, true, genuine, Gr. ἀληθινός, meaning, par excellence, red, the most glowing colour; as in Spanish *colorado* means not merely tinted, but blood-red;

Rubia tinctorum, L.

WART-CRESS, a cress with wart-shaped fruit;

Senebiera Coronopus, Poir.

WART-SPURGE, or WART-WEED, from being used to cure warts, Euphorbia helioscopia, L.

WATER AGRIMONY, Bidens tripartita, L.

WATER BETONY, Scrophularia aquatica, L.

WATER BLINKS, Montia fontana, L.

WATER-CAN, Da. *aakande*, from the shape of the seed vessel, Nuphar luteum, L.

WATER CHICKWEED, Montia fontana, L.

WATER CRESS, Nasturtium officinale, RB.

WATER CROWFOOT, Ranunculus aquatilis, L.

WATER DOCK, Rumex Hydrolapathum, L.

WATER DROPWORT, Œnanthe fistulosa, L.

WATER ELDER, Viburnum Opulus, L.

WATER FEATHERFOIL,	Hottonia palustris, L.
WATER FENNEL,	Œnanthe Phellandrium, L.
WATER FERN,	Osmunda regalis, L.
WATER FLAG,	Iris Pseudacorus, L.
WATER GERMANDER,	Teucrium Scordium, L.
WATER HEMP,	Bidens tripartita, L.
WATER HEMLOCK,	Cicuta virosa, L.
WATER HOREHOUND,	Lycopus europæus, L.
WATER HORSETAIL,	Chara, L.
WATER LILY, FRINGED-,	Villarsia nymphæoides, L.
,, ,, WHITE-,	Nymphæa alba, L.
,, ,, YELLOW-,	Nuphar luteum, L.
WATER LENTILS,	Lemna, L.
WATER MILFOIL	Myriophyllum verticillatum, L.
WATER MOSS,	Fontinalis antipyretica, L.
WATER PARSNIP,	Sium latifolium, L.
WATER PEPPER,	Polygonum Hydropiper, L.
WATER PIMPERNEL,	Veronica Beccabunga, L.
WATER PLANTAIN,	Alisma Plantago, L.
WATER PURSLANE,	Peplis Portula, L.
WATER ROCKET,	Sisymbrium sylvestre, L.
WATER SCORPION-GRASS,	Myosotis palustris, L.

WATER SOLDIER, from its sword-shaped leaves,

Stratiotes aloides, L.

| WATER SPIKE, | Potamogeton, L. |

WATER STARWORT, from its starry tufts of leaves,

Callitriche, L.

WATER THYME, a name given to a pestilent weed from North America, that chokes our canals, from some resemblance of its leaves to those of thyme,

Elodea canadensis, Rd.

WATER TORCH, in Newton's Herbal of the Bible, the reed-mace, Typha latifolia, L.

| WATER VIOLET, | Hottonia palustris, L. |
| WATER WORT, | Elatine Hydropiper, L. |

WAY BARLEY, -BENNET, or -BENT, from its growing by waysides,   Hordeum murinum, L.

WAYBREAD, the plaintain, A.S. *wegbræd*, Da. *vejbred*, G. of Ort. San. *wegbreyt*, and *wegbreidt*, a word the meaning of which is very uncertain. O. Cockayne (Leechdoms, iii. p. 347) explains it as *waybroad*. It probably meant "spread on the way." Thus the Ort. San., under *Incensaria*, says, that it is " beynahet als *wegbreidte* und wechst auch an den sandigen bergen, und breitet sich langes die erde."   Plantago major, L.

WAYFARING- or WAYFARER-TREE, from growing in hedges by the road-side, a punning name given to it by Gerarde, as implying that it is " ever on the road,"   Viburnum Lantana, L.

WAY-THORN, highway-thorn, G. *wegedorn*,   Rhamnus catharticus, L.

WEASEL-SNOUT, from the shape of the corolla, the yellow dead nettle,   Lamium Galeobdolon, Crz.

WEED, AMERICAN RIVER-, a name that, for want of a more distinctive one, is now adopted for a pestilent weed that was some years ago introduced with Canadian timber, and now infests our rivers, ponds, and canals. It has hitherto been generally known as " Babington's curse ;" a name that conveys a most unjust imputation upon a distinguished botanist, who, except drawing up an able description of the plant, has had nothing at all to do with it.   Elodea canadensis, Rd.

WEED-WIND, a corruption of *Withwind*.

WEEPING WILLOW, a tree supposed, from the resemblance of its delicate pendulous branches to long dishevelled hair, the conventional expression of grief, to be the willow of Psalm cxxxvii. 1: " By the rivers of Babylon we sat down, we *wept*. We hanged our harps upon the *willows*."   Salix babylonica, L.

WELCOME-TO-OUR-HOUSE, perhaps a quibble on its name,

Cyparissias, as meaning " Sip ere ye see us," " help your-self to the tankard, without waiting to be asked,"

Euphorbia Cyparissias, L.

WELD, WOULD, or WOOLD, Sp. *gualda*,

Reseda Luteola, L.

WELSH ONION, G. *wälsch*, foreign, being a Siberian species, and introduced into England from Germany,

Allium fistulosum, L.

WELSH POPPY, from its growing in Wales,

Meconopsis cambrica, DC.

WHARRE, a crab, W. *chwerw*, austere, bitter,

Pyrus Malus, L.

WHEAT, a term used in the first place with the meaning of *white*, wheat being, in contrast to rye, and black oats, and the black barley of Northern Asia, a white grain, A.S. *hwæte*, Go. *hvaiteis*, O.N. *hveiti*, O.H.G. *hveizi*, Lith. *kwetys*, Skr. *s'véta*, white, the initial *sv* answering, as in other cases, to a German *hu*, and Lith. *kw*,   Triticum, L.

„ BUCK-, G. *buchwaitzen*,

Polygonum Fagopyrum, L.

„ COW-, see under COW-WHEAT.

WHIN, from Fr. *guinda* or *guindoula*, M.Lat. *guindolum*, a word that generally means a kind of cherry, but in Languedoc is applied to the jujube, L. ziziphus, a name extended to other thorny and prickly shrubs (see HIP), the furze bush,   Ulex europæus, L.

„ PETTY-,   Ononis arvensis, L.

WHIN-BERRY, or WIMBERRY, the bilberry, from its growing on whins or heaths,

Vaccinium Myrtillus, L.

WHIP-TONGUE, from children using its leaves in play to draw blood from their tongues,   Galium Mollugo, L.

WHITE BEAM-TREE, a pleonasm, as A.S. *beam* means simply a tree. It is called *White Beam* from the white down on the young shoots and under surface of the leaves;

but *Beam-tree*, as it is often given, without the *White*, is a
vague and silly term. Pyrus Aria, L.

WHITE BEN, from Ar. *Behen*, Silene inflata, L.

WHITE BLOW, a name given to two of our earliest spring
flowers very conspicuous upon walls, and also called *Whit-
low-grasses*, of which *White Blow* may perhaps be a corruption. Saxifraga tridactylites, L., and Draba verna, L.

WHITE BOTTLE, from the shape of the calyx, and in distinction from the blue bottle, Silene inflata, L.

WHITE-ROOT, or -WORT, its officinal name, the Solomon's
seal, Convallaria Polygonatum, L.

WHITE ROT, from its being supposed to bane sheep,
Hydrocotyle vulgaris, L.
also, for the same reason, the butterwort,
Pinguicula vulgaris, L.

WHITE POTHERB, in distinction from the black potherb
or *olus atrum*, the lamb's lettuce,
Valerianella olitoria, L.

WHITE-THORN, Fr. *aubespine*, in distinction from the
sloe or black-thorn, from its comparatively light-coloured
rind, Cratægus Oxyacantha, L.

WHITE WILD-VINE, the white bryony, in distinction
from the black bryony, L. *vitis alba*, Plin. (l. xxiii. i. 16),
Gr. ἀμπελολευκη, Bryonia dioica, L.

WHITLOW-GRASS, a name given to two small spring
flowers from their being supposed to be the παρωνυχια of
Dioscorides, and useful in the cure of whitlows,
Saxifraga tridactylites, L., and Draba verna, L.

WHITTEN-TREE, a tree so called from its white branches;
in Berkshire, the wayfarer-tree, Viburnum Lantana, L.
but according to Gerarde (p. 1237), the water-elder,
Viburnum Opulus, L.

WHORT, or WHORTLEBERRY, the bilberry, corrupted from
L. *myrta* and *myrtillus*, the names in old vocabularies of
the myrtle-berry, a fruit largely imported in the Middle

Ages, and used in medicine and cookery, and one that the bilberry much resembles in outward appearance, the *m* being replaced with *w* as in many other instances : e.g. in *wick* from Lat. *myxa*,   Vaccinium Myrtillus, L.

WICH or WITCH ELM, see WYCH.

WICKEN-TREE, see QUICKEN.

WILDING, the crab apple, contrasted with the sweeting or cultivated sweet apple ; as in Spenser (F. Q. iii. vii. 17) :

     "Oft from the forest wildings did he bring."

                   Pyrus Malus, L.

WILL-OF-THE-WISP, from its sudden and mysterious growth by night, as if dropped by some phantom,

                   Tremella Nostoc, L.

WILLIAM, from Fr. *oeillet*, a little eye, corrupted to *Willie*, and thence to *William*, L. *ocellus*.

   ,,   SWEET-, from its scent, and partly, perhaps, in allusion to the hero of a popular ballad, "Fair Margaret and Sweet William," if this was really in existence above 300 years ago. According to an article in the Quarterly Review (No. 227), it formerly bore the name of "Sweet Saint William;" but the writer gives no reference, and probably had no authority for saying so. Bulleyn assigns the name to the wallflower.   Dianthus barbatus, L.

   ,,   WILD, the Ragged Robin,

                   Lychnis flos cuculi, L.

WILLOW, a word that seems to express a pliancy, a willingness of disposition that well accords with the character of this tree, whose branches have from time immemorial been used for wicker-work, A.S. *wilig*, L.G. *wilge*,

		Salix, L.
,,	CRACK-,	S. fragilis, L.
,,	GOATS'-,	S. caprea, L.
,,	SWEET-,	Myrica Gale, L.
,,	WEEPING-,	S. babylonica, L.
,,	WHITE-,	S. alba, L.

WILLOW-HERB, in W. Coles WILLOW-WEED, from its willow-like leaves, Epilobium, L.

„ SPIKED-, Lythrum Salicaria, L.

WILLOW-THORN, a thorny shrub with the habit of a willow, Hippophae rhamnoides, L.

WILLOW-WORT, in Sylvester's Du Bartas, p. 79, the loose-strife,

"So *willow-wort* makes wonted hate shake hands."

Lysimachia vulgaris, L.

WIND-FLOWER, from Gr. ἀνεμωνη, see ANEMONY,

Anemone, L.

WIND-ROSE, Rœmeria hybrida, DC.

WINDLE-STRAW, A.S. *windel-streow*, from *windan*, twist, and *streow*, straw, a grass whose halms are used for platting,

Agrostis Spica venti, L.

and Cynosurus cristatus, L.

WIN-BERRY, or WIMBERRY, probably a corruption of *whin-berry* rather than of *wine* and *berry*, although a fermented liquor was formerly made from this fruit, as it is in Russia to the present day, the whortleberry,

Vaccinium Myrtillus, L.

WINE-BERRY, in the Northern counties, the red-currant,

Ribes rubrum, L.

WINTER ACONITE, a plant allied to the aconites, and blowing at midwinter, Eranthis hyemalis, DC.

WINTER CHERRY, from its red cherry-like berry ripening against the winter, Physalis Alkekengi, L.

WINTER CRESS, Barbarea præcox, RB.

WINTER-GREEN, a name adopted by Turner from the German *winter-grün*, of the Ortus Sanitatis (c. 316). The Danish *winter-grönt* means the ivy, and it is probable that this latter, the ivy, is the rightful claimant of the name, as being so conspicuously green when the trees are most of them bare of leaf. Pyrola, L.

WINTER-WEED, from its being in winter the weed that spreads most, Veronica hederifolia, L.

WIRE-BENT, a bent-grass with wiry stems,
Nardus stricta, L.

WISDOM OF SURGEONS, (Anne Pratt's Wild Flowers, p. 221,) from its name *Sophia,* meaning in Greek "wisdom,"
Sisymbrium Sophia, L.

WITCHES'-BUTTER, Fris. *traal-butter,* Sw. *troll-smör,* from its buttery appearance, and unaccountably rapid growth in the night, which has given rise to a superstitious belief, still prevalent in Sweden, that witches milk the cows, and scatter about the butter on the ground,
Exidia glandulosa, Bull.

WITCH ELM, or -HAZEL, a mistaken spelling. See WYCH.

WITCHES' THIMBLE, Silene maritima, L.

WITCHEN, or QUICKEN, or WICKEN, the roan-tree, from *quycchyn,* move (Pr. Pm. p. 421), a word related to A.S. *cwic,* alive. Evelyn looking upon it as derived from *witch,* supposes it to be so called, because " it is reputed to be a preservative against fascination and evil spirits, if the boughs be stuck about the house, or used for walking staffs." It would seem in the first place to have meant the aspen, and through some mistake to have been transferred to the roan. Pyrus aucuparia, Gärt.

WITHWIND, A.S. *wiðwinde,* from *wið,* about, and *windan,* wind, Convolvulus arvensis, L.

WITHY, A.S. *wiðige, wiðie,* or *wiððe,* G. *wiede,* and etymologically identical with Du. *winde,* standing to it in the same relation as *lithy* to *linde.* Words closely related to it occur in other languages, as the Lat. *vitis* and *vimen,* Gr. ἰτέα or γιτέα, and Pers. *bid,* all derived from a root *vi,* the Skr. *wê,* and having the sense of twisting and twining, the especial use of the osier in all countries. See VINE.
Salix viminalis, L.

WOAD, or WADE, A.S. *wad,* O.S. *wode,* O.H.G. *weit,* in

Charlemagne's capitulary *waisda*, whence O.Fr. *guesde*, Fr. *guède*, and *gaide*, M.Lat. *guasdium*, *guesdium*, words derived originally from some ancient barbaric language,

Isatis tinctoria, L.

WOAD, WILD-, Reseda Luteola, L.

WOLFSBANE, wolf-poison, a plant so called because, says Gerarde (p. 822), "the hunters which seeke after woolfes, put the juice thereof into rawe flesh, which the woolfes devoure, and are killed," Aconitum Lycoctonum, L.

WOLF'S-CLAW, from the claw-like ends of the trailing stems, Lycopodium clavatum, L.

WOLF'S MILK, from the acrid qualities of its milky juice. Talbot, in Eng. Etym., suggests that the name has arisen from a confusion of Gr. λευκος, white, with λυκος, wolf; but the plant does not seem to have been called either white- or wolf's-milk by any Greek writer.

Euphorbia, L.

WOODBINE, not a bine that grows in woods, but a creeper that binds or entwines trees, in old authors called WOOD-VYNDE and WOODBINDE, A.S. *wudu-winde* and *wudu-bind*, from *wudu*, a tree, and *windan*, twine, or *bindan*, bind, G. in Tabernæm. ii. 616, *Wald-winde*, It. *Vincibosco*,

Lonicera Periclymenum, L.

It would seem in some passages to mean the bittersweet, as in Mids. N. Dr. iv. 1.

> "So doth the woodbine the sweet honeysuckle
> Gently entwist."

WOOD BLADES, -GRASS, or -RUSH,

. Luzula sylvatica, DC.

WOOD CROWFOOT, of Parkinson, the wood anemony, from its leaves resembling those of a crowfoot,

Anemone nemorosa, L.

WOOD LAUREL, L. *laureola*, dim. of *laurus*, a name under which all evergreen shrubs were once included,

. Daphne Laureola, L.

Wood Lily, the lily of the valley,
                    Convallaria majalis, L.
Wood Nightshade, the bittersweet,
                    Solanum Dulcamara, L.
Wood Nut,   Corylus Avellana, L.
Wood Pea, from its small pea-like tubers, and its usually growing in woods,   Vicia Orobus, DC.
Wood Reed, in distinction from the pool-reed,
                    Calamagrostis epigeios, L.
Wood-roof, -rofe, -row, -rowel, or -ruff, and agreeably to an old distich :

"Double U, double O, double D, E,
R, O, double U, double F, E,"

Woodderowffe, A.S. *wude-rofe*, from Fr. *roue*, a wheel, and its dim. *rouelle*, a little wheel or rowel, the leaves being set upon the stem in verticils that resemble the large rowels of ancient spurs.   This is one among several other words that we find to have been adopted into Anglo-Saxon from the French, an occurrence a good deal more frequent than philologists seem to be aware, who, looking upon the former as a pure Germanic language, would trace its vocabulary too exclusively to native roots.
                    Asperula odorata, L.
Wood Rush, or -Grass,   Luzula sylvatica, DC.
Wood Sage,   Teucrium Scorodonia, L.
Wood Sorrel, or -Sowr,   Oxalis Acetosella, L.
Wood Spurge,   Euphorbia amygdaloides, L.
Wood Vetch,   Vicia Orobus, DC.
Wood Vine,   Bryonia dioica, L.
Wood-waxen, A.S. *wudu-weaxe*, which would seem to mean "wood-grown," a word very inapplicable to a plant that is always found in open meadows.   It is most probably a corruption of some German name meaning " woad-plant," *waud-gewächse*.   It is called in Sloane MS. 1571, 3,'
Wodewex,   Genista tinctoria, L.

WOOLD, a dyer's term, see WELD.

WOOLLEN, the mullein, from its woolly leaf, Fl. *wolle-kruydt*, G. of Ort. San. c. 110, *Wulkraut*,

Verbascum Thapsus, L.

WORM-GRASS, from its vermifuge qualities,

Sedum album, L.

WORM-SEED, from its reputed vermifuge qualities,

Erysimum cheiranthoides, L.

WORMWOOD, a word corrupted from A.S. *wermod*, G. *wer-muth*, O.H.G. *werimuota*, O.S. *weremede*, words which seem to be compounded with G. *wehren*, A.S. *werian*, keep off, *mod* or *made*, maggot, but which, by an accidental coincidence of sound, have been understood as though the first syllable were *worm*. L. Diefenbach would prefer to derive it from a Celtic root that means "bitter," Wel. *chwerw*, Corn. *wherow*. Be its origin what it may, it was understood in the Middle Ages as meaning a herb obnoxious to maggots, and used to preserve things from them, and was also given as an anthelmintic or worm medicine.

Artemisia Absinthium, L.

WORTS, see WHORTLEBERRY,

Vaccinium Myrtillus, L.

WORTS, in Chaucer a general name for cultivated plants, A.S. *wyrt*, Go. *aurts*, Skr. *vridh*, grow ; but in Shakspeare (M.W.W. act i. sc. 1), and in Lupton's Notable Things, more especially cabbage, being an abbreviation of *cole-worts*,   Brassica, L.

WOUNDWORT, from its soft downy leaves having been used instead of lint for dressing wounds,

Stachys Germanica, L.

also, for the same reason,   Anthyllis vulneraria, L.

  ,,   CLOWN'S, see under CLOWN.

  ,,   KNIGHT'S WATER-, a plant that from its sword-shaped leaves was supposed, on the doctrine of signatures, to heal sword-wounds,   Stratiotes aloides, L.

WRACK, seaweed thrown ashore, from a Norse or Frisian word connected with Da. *vrage*, reject, Du. *wraken*,

    „   GRASS-, a sea-plant with long grass-like leaves,
Zostera marina, L.

WYCH-ELM, an elm so named from its wood having been used to make the chests called in old writers *wyches, hucches,* or *whycches*, from Fr. *huche*, A.S. *hwœcce*, a term applied by Sir John Mandeville (c. viii.) to the Ark of the Testimony ; and in a poem called " Cleanness," edited by Dr. R. Morris, to Noah's Ark :

> " And alle woned in þe *whicche* þe wylde and þe tame ;"

but more generally to the boxes used for keeping provisions, as in Hazlitt's Early Popular Poetry, p. 210 :

> " His hall rofe was full of bacon flytches,
> The chambre charged was with *wyches*
> Full of egges, butter, and chese."

Ulmus montana, L.

WYCH-HAZEL, from the resemblance of its leaf to that of the hazel, the wych-elm,   Ulmus montana, L.

YARR, abbreviated from *yarrow*, and applied to a very different plant, the spurry, from both having been confused under the name of *milfoil*,   Spergula arvensis, L.

YARROW, the milfoil, A.S. *gearwe*, L.Ger. *geruwe*, O.H.G. *garawa*, O.Fris, *kerva*, G. *garbe*, a word that seems to have been properly the name of the vervain, *hiera-botane*, the *gerebotanon* of Apuleius, c. iii., from Gr. ἰερα βοτανη, holy herb, with which and with the betony we learn from a couplet in Macer, c. 58, that it was associated in its vulnerary and other supposed virtues :

> " Herbam, cui nomen *foliis* de *mille* dedere,
> Betonicamque pari verbenæ pondere junge."

The initial *hi* of Greek words has in the Germanic languages been usually replaced with *y* or *j*, and thus, as *Hieronymus* and *Hierosolyma* have become *Jerom* and *Jerusalem*, so *hiera* has become *yarrow*.   Achillæa Millefolium, L.

YEAST-PLANT,   Penicillium glaucum, Ber.

YELLOW ARCHANGEL, see ARCHANGEL,

Lamium Galeobdolon, Crz.

YELLOW BIRDSNEST, in contrast to the wild carrot, that was also called *Birdsnest*,   Monotropa Hypopitys, L.

YELLOW BUGLE,   Ajuga Chamæpitys, L.

YELLOW CRESS,   Barbarea præcox, RB.

YELLOW LOOSESTRIFE,   Lysimachia vulgaris, L.

YELLOW OX-EYE,   Chrysanthemum segetum, L.

YELLOW PIMPERNEL,   Lysimachia nemorum, L.

YELLOW RATTLE,   Rhinanthus Crista galli, L.

YELLOW ROCKET,   Barbarea vulgaris, RB.

YELLOW-WEED, a weed or dye-plant used for dyeing yellow, the term *weed* being here, as in green-weed, red-weed, etc., not the A.S. *weod*, but the Du. *weed*, G. *waid*, the weld,   Reseda Luteola, L.

YELLOW-WORT,   Chlora perfoliata, L.

YEVERING BELLS, L. *tintinnabulum terræ*, from the resemblance of its flowers to little bells hung one above the other to be struck with a hammer. *Yevering* is usually spelt *yethering*, from Scotch *yether*, beat. Pyrola secunda, L.

YEW, or YEUGH, in old authors variously spelt EWGH, UGH, EWE, and U, A.S. *iw*, O.H.G. *iwa*, G. *eibe*, Sp. and Port. *iva*, F. *if*, W. *yw*, from M.Lat. *ivus, iva*, or *iua*, a name applied to several different plants, and of uncertain derivation. Some of the dictionaries allege for it a Celtic *iw*, green, but there does not appear to be any such word. It seems to be an abbreviation of *aiuga*, a misspelling of L. *abiga*, a plant mentioned by Pliny (b. xxiv. ch. 20), as being the same as the Gr. χαμαιπιτυς, and called so from its causing abortion. Under this name *iua* we find the *yew* so inextricably mixed up with the *ivy*, that, as dissimilar as are the two trees, there can be no doubt that their names are in their origin identical. How they came to be attached to them both, is the difficulty. Apuleius (ch. 26), speaking of chamæpitys, says, " Græci chamæpityn, Itali *abigam*, alii cupressum nigram vocant." Brunsfels too says of the chamæpitys (b. i. p. 161), " Ego autem cipressen existimavi." The yew seems to have been taken for this black cypress, and in this way to have acquired the terms *abiga* and *ajuga*, and *iua* and *iva*. But we learn from Parkinson (Th. Bot. p. 284) that the forget-me-not, a weed of corn-fields, was " called in English Ground pine, and *Ground ivie*, after the Latin word *Iva*." This term *Ground ivy* was assigned by others to another small labiate plant, (Nepeta Glechoma, B.) which was formerly called Hedera terrestris, and *ivy* regarded as the equivalent of *hedera*, and subsequently transferred to the Hedera helix, our present *Ivy*. The origin of *Ajuga* seems to have been a mere error of the copyist in transcribing the passage from the 24th book of Pliny. For, as distinct as are *abiga* and *ajuga* in our modern print, the *b* of *abiga* might be written so as to look like a *v* or *u*, and the word made to appear *auiga*, which, if the *i* were not dotted, might be as easily read *aiuga* as *auiga*. See Ivy. Thus by a train of blunders, Pliny's *abiga* becomes *ajuga*, and *ajuga iua* or *iva*. This *abiga*, (*ajuga*, or *iua*) was, as Pliny tells us, the same as the Greek *chamæpitys*. The yew-tree gets the name of *chamæpitys* through a remark made by Apuleius, and thereby, as its synonym, that of *iua* or *iva*. The ground-pine, from its terebinthinate odour, also gets the name of *chamæpitys*, and thereby, as its synonym, that of *iua* or *iva*. But from *chamæ*- being equivalent to *terrestris*, this name *iua* or *iva* passes over to a weed called, from the shape of its leaf and creeping habit, *hedera terrestris*, and the equivocal word *hedera* conveys it to the shrub which thus gets the name of *ivy*.

Taxus baccata, L.

YOKE-ELM, the hornbeam, from yokes being made of it, Gr. ζυγια,   Carpinus Betulus, L.

YORKSHIRE SANICLE, the butterwort, from being, for its healing qualities, called by Bauhin (Pin. 243) *Sanicula*, and "growing so plentifully in Yorkshire," as Parkinson tells us (Th. Bot. p. 534),   Pinguicula vulgaris, L.

YORNUT, YERNUT, or YENNUT, in the Northern counties, a dialectic pronunciation of *Earthnut*, Da. *jord-nöd*,   Bunium flexuosum, With.

YOUTHWORT, A.S. *eowð*, a flock, and *rotian*, rot, mistaken for *wort*, so called from its being supposed to bane sheep, the red-rot,   Drosera rotundifolia, L.

SYSTEMATIC NAMES OF BRITISH PLANTS,

WITH THE OMISSION OF SUCH AS HAVE NO POPULAR EQUIVALENT.

ACER CAMPESTRE, L. Maple. Maser-tree.
,, PSEUDOPLATANUS, L. Sycamore. Mock Plane.
ACERAS ANTHROPOPHORA, R.B. Man-, or Green Man Orchis.
ACHILLÆA MILLEFOLIUM, L. Yarrow. Nosebleed. Milfoil.
Sanguinary.
,, PTARMICA, L. Sneezewort. Goose-tongue.
,, AGERATUM, L. Maudlin.
ACONITUM NAPELLUS, L. Monkshood. Wolfsbane. Aconite.
Friar's cap.
ACORUS CALAMUS, L. Sweet Flag. Sweet Sedge.
ACTÆA SPICATA, L. Baneberry. Herb Christopher.
ACTINOCARPUS, see DAMASONIUM.
ADIANTUM CAPILLUS, L. Maidenhair. Capillaire. Venus' Hair.
ADONIS AUTUMNALIS, L. Pheasant's eye. Red Mayd-weed. Rose-
a-ruby. Red Morocco.
ADOXA MOSCHATELLINA, L. Moscatel.
ÆGOPODIUM PODAGRARIA, L. Goutweed. Ashweed. Herb Gerard.
ÆSCULUS HIPPOCASTANUM, L. Horse Chesnut.
ÆTHUSA CYNAPIUM, L. Fool's Parsley. Asses' Parsley. Dog's
Parsley.
AGARICUS OREADES, Bolt. Champignon. Pixie-stools.
,, MUSCARIUS, L. Bug Agaric. Flybane.
,, CAMPESTRIS, L. Mushroom.
,, ARVENSIS, Sch. Horse Mushroom.
AGRIMONIA EUPATORIA, L. Agrimony. Egremoine.
AGROSTEMMA, see LYCHNIS.
AGROSTIS ALBA, L. Fiorin.
,, ,, var. stolonifera. Knot-grass.

AGROSTIS SPICA VENTI, L. Windlestraw.

AIRA CÆSPITOSA, L. Tussack-grass.

,, CARYOPHYLLEA, L. Hair-grass.

AJUGA CHAMÆPITYS, L. Ground-pine. Forget-me-not. Herb Ivy. Gout Ivy. Field Cypress.

,, REPTANS, L. Bugle.

ALARIA ESCULENTA, Lam. Honeyware.

ALCHEMILLA ARVENSIS, L. Parsley-piert. Breakstone. Percepier.

,, VULGARIS, L. Lady's mantle. Lion's foot. Padelion. Syndaw.

ALISMA PLANTAGO, L. Water Plantain.

ALLIARIA OFFICINALIS, DC. Jack-by-the-hedge. Sauce-alone. Garlick Mustard. Garlick-wort.

ALLIUM ASCALONICUM, L. Shallot. Scallion. Cibbols.

,, CEPA, L. Onion.

,, SCORODOPRASUM, L. Rocambole.

,, SCHŒNOPRASUM, L. Chives.

,, VINEALE, L. Crow Garlick.

,, URSINUM, L. Ramsons. Bear's Garlick. Buckrams.

,, FISTULOSUM, L. Welsh Onion.

,, PORRUM, L. Leek. Purret.

,, SATIVUM, L. Garlick. Poor-man's treacle. Churl's treacle.

ALLOSORUS CRISPUS, Ber. Parsley-fern.

ALNUS GLUTINOSUS, L. Alder.

ALOPECURUS AGRESTIS, L. Black Bent. Mouse-tail Grass. Hunger Grass.

,, PRATENSIS, L. Meadow Foxtail.

ALTHÆA OFFICINALIS, L. Marsh Mallow. Hock-herb.

,, ROSEA, L. Hollihock.

ALYSSUM MARITIMUM, L. Sweet Alison.

,, SAXATILE, L. Yellow Alison.

AMARANTUS CAUDATUS, L. Florimer. Love-lies-bleeding. Amaranth. Thrumwort. Velvet-flower.

,, HYPOCHONDRIACUS, L. Prince's feather.

AMBROSIA, L. Ambrose.

AMMI MAJUS, L. Bull-wort. Herb William. Bishop's weed.

AMYGDALUS PERSICA, W. Peach. Nectarine.

ANACHARIS ALSINASTRUM, Bab. See ELODEA CANADENSIS, Rd.

ANACYCLUS PYRETHRUM, DC. Pellitory of Spain.
ANAGALLIS ARVENSIS, L. Red Pimpernell. Poor-man's Weather-
glass.
ANCHUSA OFFICINALIS, L. Alkanet. Bugloss.
ANDROMEDA POLIFOLIA, L. Moor-wort. Marsh Rosemary. Marsh
Holyrose.
ANEMONE NEMOROSA, L. Wind-flower. Wood Crowfoot. Wild
Anemony.
  ,,   PULSATILLA, L. Flaw-flower. Pasque-flower.
  ,,   HEPATICA, L. Noble Liverwort. Hepatica. Liverleaf.
ANETHUM GRAVEOLENS, L. Dill-seed. Anet.
ANGELICA ARCHANGELICA, L. Archangel.
  ,,   SYLVESTRIS, L. Holy Ghost.
ANTHEMIS NOBILIS, L. Chamomile.
  ,,   COTULA, L. Maydweed. Dog's Fennel. Mather.
ANTHOXANTHUM ODORATUM, L. Sweet-scented Vernal-grass.
ANTHRISCUS SYLVESTRIS, L. Hare's Parsley.
ANTHYLLIS VULNERARIA, L. Lady's Fingers. Kidney Vetch.
Lamb's-toe.
ANTIRRHINUM MAJUS, L. Snapdragon. Calves snout. Lion's snap.
APIUM GRAVEOLENS, L. Celery. Smallage. Marsh Parsley.
AQUILEGIA VULGARIS, L. Columbine. Culverwort.
ARABIS PERFOLIATA, Lam. Tower Mustard.
  ,,   TURRITA, L. Tower Cress.
  ,,   THALIANA, L. Codded Mouse-ear. Wall Cress.
  ,,   STRICTA, Huds. Bristol Rock Cress.
ARBUTUS UNEDO, L. Strawberry-tree.
ARCTIUM LAPPA, L. Burdock. Hardock. Hurr-burr.
ARCTOSTAPHYLOS UVA URSI, Spr. Bearberry. Mealberry.
ARENARIA, L. Sand-wort, or -weed.
ARISTOLOCHIA CLEMATITIS, L. Birthwort.
ARMERIA VULGARIS, W. Thrift. Lady's cushion. Sea Gilliflower.
Cushion Pink.
ARTEMISIA ABSINTHIUM, L. Wormwood.
  ,,   DRACUNCULUS, L. Tarragon.
  ,,   VULGARIS, L. Mugwort. Motherwort.
  ,,   ABROTANUM, L. Southernwood. Boy's love. Lad's
love. Old Man. Averoyne.

ARUM MACULATUM, L. Cuckoo-pint. Lords and Ladies. Wake-pintle. Wake Robin. Aaron. Bloody-man's-finger. Calves-foot. Rampe. Starch-wort.

ARUNDO PHRAGMITES, L. Reed. Pole-reed. Spires.

ASARUM EUROPÆUM, L. Asarabacca. Fole-foot. Hazel-wort.

ASPARAGUS OFFICINALIS, L. Sparrow-grass. Sperage.

ASPERUGO PROCUMBENS, L. German Madwort.

ASPERULA CYNANCHICA, L. Squinancy-wort.

,, ODORATA, L. Woodroof.

ASPIDIUM LONCHITIS, Sw. Holly-fern.

,, ACULEATUM, Sw. Prickly Shield-fern.

,, THELYPTERIS, Sw. Marsh-fern.

,, OREOPTERIS, Sw. Sweet-fern. Mountain-fern.

,, FILIX MAS, Sw. Male-fern.

ASPLENIUM, L. Spleenwort. Miltwaste.

,, FILIX FŒMINA, Bern. Lady-fern.

,, MARINUM, L. Sea Spleenwort.

,, TRICHOMANES, L. Black Maidenhair.

,, RUTA MURARIA, L. Wall Rue. Tent-wort.

ASTER TRIPOLIUM, L. Sharewort. Sea Starwort.

,, TRADESCANTI, L. Michaelmas Daisy.

ASTRAGALUS GLYCYPHYLLOS, L. Liquorice Vetch.

ATRIPLEX HORTENSIS, L. Orache.

,, PORTULACOIDES, L. Sea Purslane.

,, PATULA, L. Delt Orach. Lamb's Quarters. Fat hen.

ATROPA BELLADONNA, L. Deadly Nightshade. Dwale. Death's-herb. Great Morel.

AVENA SATIVA, L. Oat. Haver.

,, NUDA, L. Pill-corn.

,, FATUA, L. Wild Oats. Drake.

BALLOTA NIGRA, L. Black Horehound.

BALSAMITA VULGARIS, L. Alecost. Maudlin. Costmary.

BARBAREA PRÆCOX, R.B. Belleisle Cress.

,, VULGARIS, R.B. Winter Cress. Yellow Rocket. St. Barbara's Cress. Land Cress.

BARTSIA ALPINA, L. Poly-mountain.

,, ODONTITES, L. Eyebright Cow-wheat.

BELLIS PERENNIS, L. Daisy. Bruisewort. Herb Margaret.
Marguerite.
BERBERIS VULGARIS, L. Barberry. Pipperidge.
BETA MARITIMA, L. Beet. Mangel-wurzel.
BETULA ALBA, L. Birch.
BIDENS CERNUA, L. Nodding Bur Marigold.
,, TRIPARTITA, L. Trifid Bur Marigold. Water Agrimony.
Water Hemp.
BLECHNUM BOREALE, Sw. Hard-fern.
BOLETUS, L. Canker.
BORAGO OFFICINALIS, L. Borage.
BOTRYCHIUM LUNARIA, Sw. Moonwort. Lunarie. Plantage.
BRASSICA OLERACEA, L. vars. Cabbage. Cauliflower. Broccoli.
Cale. Savoy. Kohl-rabi. Bore-cole.
,, CAMPESTRIS, L. var. Rapa, Rape. Coltza. Mype.
,, ,, var. Napus. Turnip. Knolles. Navew.
Rutabaga. Swede.
,, ALBA, Boiss. Mustard.
,, SINAPISTRUM, Boiss. Charlock. Wild Mustard. Chedlock.
,, NIGRA, Boiss. Black Mustard. Senvy.
,, TENUIFOLIA, Boiss. Wall Rocket.
BRIZA MEDIA, L. Quaking-grass. Wagwants. Lady's hair.
Maidenhair-grass. Shaker. Pearl-grass.
BROMUS MOLLIS, L. Lobgrass. Oatgrass.
,, STERILIS, L. Drake.
BRYONIA DIOICA, L. White Bryony. White Wild Vine.
BUNIUM FLEXUOSUM, W. Arnut. Yor-nut. Jur-nut. Pig-nut.
Mandrake. Tetter-berry. Hog-nut. Earth-nut.
St. Anthony's-nut.
BUPLEURUM ROTUNDIFOLIUM, L. Thorowax. Hare's ear.
BUTOMUS UMBELLATUS, L. Flowering Rush.
BUXUS SEMPERVIRENS, L. Box.

CAKILE MARITIMA, L. Sea Rocket.
CALAMAGROSTIS EPIGEIOS, Roth. Wood-reed.
CALAMINTHA CLINOPODIUM, Benth. Stone Basil. Field Basil.
Horse Thyme.
,, ACINOS, Clair. Basil Thyme.

CALENDULA OFFICINALIS, L. Marigold. Golde. Gools. Gowan. Ruddes. Marybuds.

CALLITRICHE AQUATICA, Sm. Star-grass.

CALTHA PALUSTRIS, L. Marsh Marigold. Brave Bassinets. Boots. Meadow Bouts. Mare-blobs.

CAMELINA SATIVA, L. Gold of pleasure. Cheet. Oilseed.

CAMPANULA RAPUNCULUS, L. Rampion. Coventry Rapes.

,, ROTUNDIFOLIA, L. Harebell. Lady's Thimble. Witches' Thimble.

,, HYBRIDA, L. Venus' Looking-glass. Lady's Looking-glass. Corn Violet.

,, TRACHELIUM, L. Canterbury Bells. Throat-wort. Hask-wort. Mercury's Violet. Mariet. Coventry Bells.

CANNABIS SATIVA, L. Hemp. Gallow-grass. Neckweed.

CANTHARELLUS CIBARIUS, Fr. Chantarelle.

CAPSELLA BURSA PASTORIS, L. Shepherd's pouch. Casse-weed. Clappedepouch. Toy-wort. Pickpurse. Poor-man's Parmacetty.

CARDAMINE AMARA, L. Bitter Cress.

,, PRATENSIS, L. Lady's smock. Cuckoo-flower. Meadow Cress. Spinks.

CARDUUS BENEDICTUS, L. Blessed Thistle.

,, ERIOPHORUS, L. Cotton Thistle. Friar's crown.

,, HETEROPHYLLUS, L. Melancholy Thistle.

,, LANCEOLATUS, L. Spear Thistle. Bur Thistle.

,, MARIANUS, L. Milk Thistle. Lady's Thistle.

,, NUTANS, L. Musk Thistle. Scotch Thistle.

CAREX, L. Sedge.

,, PANICULATA, L. Hassocks.

,, ARENARIA, L. Stare.

,, PANICEA, L. Carnation-grass.

CARLINA VULGARIS, L. Carline Thistle.

CARPINUS BETULUS, L. Hornbeam. Hurst Beech. Hard-beam. Yoke Elm.

CARTHAMUS TINCTORUM, L. Safflower.

CARUM CARUI, L. Carraway.

,, BULBOCASTANUM, L. Earth-nut. Pig-nut. Arnut.

CAUCALIS ANTHRISCUS, Huds. Hedge Parsley. Hemlock Chervil. Rough Cicely.

,, DAUCOIDES, L. Bur Parsley. Hedgehog Parsley. Hen's foot.

CENOMYCE PYXIDATA, Ach. Cup Moss.

CENTAUREA CYANUS, L. Bluebottle. Corn-flower. Blue Blaw. Hurt-sickle.

,, NIGRA, L. Knapweed. Horse-knob. Hard-head. Mat-fellon. Bullweed. Churl's head. Loggerhead.

,, CALCITRAPA, L. Caltrop.

,, SOLSTITIALIS, L. St. Barnaby's Thistle. Star Thistle.

CENTRANTHUS RUBENS, DC. Red Valerian.

CENTUNCULUS MINIMUS, L. Chaff-weed.

CERASTIUM VULGARE, L. Mouse-ear Chickweed.

CERATOPHYLLUM, L. Hornwort.

CERCIS SILIQUASTRUM, L. Judas-tree.

CETERACH OFFICINARUM, Willd. Ceterach. Scaly-fern. Finger-fern.

CETRARIA ISLANDICA, Ach. Iceland Moss.

CHÆROPHYLLUM SYLVESTRE, L. Cow Parsley. Wild Cicely.

CHARA, L. Water Horsetail. Stone-wort.

CHEIRANTHUS CHEIRI, L. Wall-flower. Bleeding-heart. Bloody warrior. Wild Cheir. Chevisaunce.

CHELIDONIUM MAJUS, L. Celandine. Swallow-wort. Tetter-wort.

CHENOPODIUM, L. Goosefoot.

,, ALBUM, L. Frostblite.

,, BONUS HENRICUS, L. Allgood. Good King Henry. Blite. English Mercury.

,, POLYSPERMUM, L. Allseed.

,, RUBRUM, L. Pig-weed. Sowbane.

,, VULVARIA, L. Notch-weed. Dog's Orach.

,, AMBROSIOIDES, L. Oak of Cappadocia. Oak of Jerusalem.

,, BOTRYS, L. Ambrose.

CHERLERIA SEDOIDES, L. Cyphel.

CHLORA PERFOLIATA, L. More Centory. Yellow-wort.

CHONDRUS CRISPUS, Lyn. Carrageen Moss. Irish Moss.

CHRYSANTHEMUM LEUCANTHEMUM, L. Moon-wort. Ox-eye. Moon Daisy. Maudlin-wort. Midsummer Daisy.

CHRYSANTHEMUM SEGETUM, L. Bigold. Boodle. Goldins. Gools.
Ruddes. Yellow Ox-eye. Corn Marigold.

,, PARTHENIUM, L. Feverfew.

CHRYSOCOMA LINOSYRIS, L. Goldilocks.

CHRYSOSPLENIUM, L. Golden Saxifrage.

CICER ARIETINUM, L. Garavance. Gram. Chick pea.

CICHORIUM INTYBUS, L. Succory. Chicory.

,, ENDIVIA, L. Endive.

CICUTA VIROSA, L. Water Hemlock. Cowbane.

CIRCÆA LUTETIANA, L. Enchanter's Nightshade.

CLADIUM MARISCUS, L. Twig-rush.

CLADONIA RANGIFERINA, Hff. Reindeer Moss.

CLEMATIS VITALBA, L. Virgin's bower. Lady's bower. Old-
man's-beard. Traveller's joy. Bind-with. Hedge-
vine. Love. Smoke-wood. Climbers.

COCHLEARIA OFFICINALIS, L. Scurvy-grass. Spoonwort.

,, ARMORACIA, L. Horse Radish.

COLCHICUM AUTUMNALE, L. Meadow Saffron. Naked Ladies.
Upstart.

COMARUM, see POTENTILLA.

CONFERVA ÆGAGROPILA, L. Moorballs.

CONIUM MACULATUM, L. Hemlock. Herb Bennett.

CONVALLARIA MAJALIS, L. Lily of the Valley. Liry-confancy.
May-lily. Lily-convally.

,, POLYGONATUM, L. Solomon's seal. Ladder-to-heaven.
Lady's seal. Seal-wort. White-root.

CONVOLVULUS ARVENSIS, L. Bindweed. Bearbind. Withwind.
Cornbind.

,, SEPIUM, L. Hedge-bells. Lady's nightcap. Campanelle.

,, SOLDANELLA, L. Sea-bells, -Bindweed, or -Withwind.

CORALLORHIZA INNATA, RB. Coral-root.

CORIANDRUM SATIVUM, L. Coriander. Col.

CORNUS SANGUINEA, L. Dogwood. Gadrise. Dog-cherry.

,, SUECICA, L. Dwarf Honeysuckle.

CORRIGIOLA LITTORALIS, L. Strapwort.

CORYDALIS TUBEROSA, DC. Holewort. Hollowort.

CORYLUS AVELLANA, L. Hazel. Stocknut. Filbert. Cobnut.

COTYLEDON UMBILICUS, L.   Navel-wort.   Kidney-wort.   Hip-
    wort.   Lady's navel.   Wall Pennywort.
CRAMBE MARITIMA, L.   Sea-kale.   Sea Cabbage.
CRATÆGUS OXYACANTHA, L.   Hawthorn.   Quickset.   White-thorn.
    May.   Albespyne.
CREPIS, L.   Hawksbeard.
CRITHMUM MARITIMUM, L.   Samphire.
CROCUS SATIVUS, L.   Saffron.
CUCUMIS MELO, L.   Melon.
    ,,   SATIVUS, L.   Cucumber.
CUCURBITA PEPO, L.   Gourd.   Pumpkin.
    ,,   OVIFERA, W.   Vegetable marrow.
CUMINUM CYMINUM, L.   Cummin.
CUSCUTA EUROPÆA, L.   Dodder.   Lady's laces.   Bride's laces.
    Hell-weed.   Devil's guts.   Strangle-tare.
CYCLAMEN EUROPÆUM, L.   Sowbread.   Cyclamen.
CYNARA SCOLYMUS, L.   Artichoke.
    ,,   CARDUNCULUS, L.   Cardoon.
CYNODON DACTYLON, L.   Doob.   Dog's tooth.
CYNOGLOSSUM OFFICINALE, L.   Hound's tongue.   Dog's tongue.
CYNOSURUS CRISTATUS, L.   Dogstail.
    ,,   ECHINATUS, L.   Cock's comb grass.
CYPERUS LONGUS, L.   Cypress-root.   Sweet Cypress.   Galangale.
CYPRIPEDIUM CALCEOLUS, L.   Lady's slipper.
CYSTOPTERIS FRAGILIS, Bern.   Bladder-fern.
CYTISUS LABURNUM, L.   Laburnum.   Golden chain.

DACTYLIS GLOMERATA, L.   Orchard-grass.   Dew-grass.
DAMASONIUM STELLATUM, P.   Star-fruit.
DAPHNE LAUREOLA, L.   Spurge-, or Wood-, or Copse-Laurel.
    Lowry.   Daphne.
    ,,   MEZEREON, L.   Mezereon.   Spurge Olive.
DATURA STRAMONIUM, L.   Thorn-apple.   Dewtry.
DAUCUS CAROTA, L.   Carrot.   Bee's nest.   Dauke.
DELPHINIUM, L.   Larkspur.   Knight's spurs.
    ,,   GRANDIFLORUM, L.   Bee Larkspur.
    ,,   STAPHISAGRIA, L.   Stavesacre.
    ,,   CONSOLIDA, L.   Consound.

DENTARIA BULBIFERA, L. Tooth Violet. Coral-wort.
DIANTHUS, L. Pink.
,, ARMERIA, L. Deptford Pink.
,, DELTOIDES, L. Maiden Pink. Meadow Pink.
,, CÆSIUS, L. Cheddar Pink. Cliff Pink.
,, BARBATUS, L. Sweet William. Sweet John. Tol-
meiner.
,, PROLIFER, L. Childing Pink.
,, CARYOPHYLLUS, L. Carnation. Clove Pink. Gilli-
flower. Piggesnie. Sops-in-wine.
DIGITALIS PURPUREA, L. Foxglove. Finger flower.
DIGRAPHIS ARUNDINACEA, P.B. Lady's garters. French-grass.
Ribbon-grass.
DIOTIS MARITIMA, Dsf. Sea Cudweed.
DIPSACUS FULLONUM, L. Fuller's teasel.
,, PILOSUS, L. Shepherd's staff, or -rod.
,, SYLVESTRIS, L. Teasel. Venus' bason.
DRABA VERNA, L. Whitlow-grass. Nail-wort. White Blow.
DROSERA ROTUNDIFOLIA, L. Sundew. Lustwort. Youthwort.
DRYAS OCTOPETALA, L. Mountain Avens.

ECHIUM VULGARE, L. Viper's-bugloss.
ELATINE HYDROPIPER, L. Water Pepper. Water-wort.
ELODEA CANADENSIS, Rd. American River weed. Water Thyme.
ELYMUS ARENARIUS, L. Lyme-grass.
EMPETRUM NIGRUM, L. Crowberry. Crakeberry.
EPILOBIUM HIRSUTUM, L. Codlins and Cream. Willow-herb.
,, ANGUSTIFOLIUM, L. French willow. Persian willow.
Rose-bay.
EPIMEDIUM ALPINUM, L. Barren-wort.
EPIPACTIS, RB. Helleborine.
EQUISETUM ARVENSE, L. Bottle-brush.
,, HYEMALE, L. Dutch-rush. Shave-grass. Pewter-wort.
Scouring-rush.
,, LIMOSUM, L. Paddock-pipes. Toad-pipes.
,, TELMATEJA, Ehr. Great Horsetail.
ERANTHIS HYEMALIS, DC. Winter Aconite.
ERICA TETRALIX, L. Cross-leaved Heath.

ERICA CINEREA, L. Grey Heath. Scotch Heath.

,,   VULGARIS, L. Ling. Heath. Grigg.

,,   VAGANS, L. Cornish Heath.

ERIGERON ACRE, L. Blue Fleabane.

ERIOCAULON SEPTANGULARE, L. Pipe-wort.

ERIOPHORUM, L. Cotton-rush.

,,   VAGINATUM, L. Hare's-tail-rush. Moss-crops.

ERISIPHE, DC. Mildew.

ERODIUM MOSCHATUM, L'Her. Heron's bill. Muscovy. Musk.
    Pink-, Powk-, or Pick-needle.

ERVUM LENS, L. Lentil. Tills.

,,   ERVILIA, L. Ers. Pigeon's pea.

ERYNGIUM MARITIMUM, L. Eryngo. Sea Holly.

ERYSIMUM CHEIRANTHOIDES, L. Treacle Mustard.

ERYTHRÆA CENTAURIUM, L. Lesser Centaury. Earth-gall. Christ's
    ladder.

EUPATORIUM CANNABINUM, L. Hemp Agrimony. Holy rope.

EUPHORBIA HELIOSCOPIA, L. Sun-spurge. Turnsole. Wart-
    weed. Devil's milk. Cat's milk. Littlegood. Churn-
    staff.

,,   LATHYRIS, L. Wild Capers.

,,   CYPARISSIAS, L. Welcome-to-our-house.

EUPHRASIA OFFICINALIS, L. Eyebright. Euphrasy.

EVONYMUS EUROPÆUS, L. Spindle-tree. Prick-wood. Skewer-
    wood. Gadrise. Louse-berry-tree.

EXIDIA GLANDULOSA, B. Witches butter.

,,   AURICULA JUDÆ, Fr. Jew's ears.

FAGUS SYLVATICA, L. Beech. Buck-mast.

FESTUCA PRATENSIS, L. Fescue-grass.

FICUS CARICA, L. Fig-tree.

FILAGO, see GNAPHALIUM.

FILIX, Fern.

FÆNICULUM VULGARE, Gärt. Fennel.

FONTINALIS ANTIPYRETICA, L. Water Moss.

FRAGARIA VESCA, L. Strawberry.

FRANKENIA LÆVIS, L. Sea Heath.

FRAXINUS EXCELSIOR, L. Ash.

Fritillaria Meleagris, L. Fritillary. Guinea hen. Checkered lily. Snake's head.

Fucus nodosus, L. Kelpware. Tang. Knob-tang.

,, natans, Turn. See Sargassum.

Fumaria officinalis, L. Fumitory. Earth-smoke.

Fungus, L. Mushroom. Toadstool. Paddock-stool.

Gagea lutea, Ker. Yellow Star of Bethlehem.

Galanthus nivalis, L. Snowdrop. Fair Maids of February. Purification-flower.

Galeopsis Ladanum, L. Red Hemp-nettle. Iron-wort.

,, Tetrahit, L. Hemp-nettle. Bee-nettle.

Galium Cruciata, Scop. Crosswort. Maywort. Golden Mug-weet.

,, Mollugo, L. Whip-tongue. White Bedstraw.

,, verum, L. Lady's Bedstraw. Maid's hair. Petty Mugget. Cheese Rennet.

,, Aparine, L. Cleavers. Cliders. Goosegrass. Goosebill. Harif. Goose-heiriffe. Loveman. Beggar's lice. Scratch-weed. Catch-weed. Grip-grass.

Gastridium lendigerum, P.B. Nit-grass.

Geaster, B. Earth-star.

Genista tinctoria, L, Base-broom. Dyer's Green-weed. Wood-waxen.

,, anglica, L. Petty Whin. Needle Furze. Moor-, or Moss-Whin.

Gentiana, L. Bitterwort. Felwort. Gentian.

,, Pneumonanthe, L. Autumn bells. Calathian Violet. Lung-flower.

Geranium pratense, L. Meadow Cranesbill. Crowfoot Cranes-bill.

,, Robertianum, L. Herb Robert. Red-shanks.

,, molle, L. Dove's foot.

,, columbinum, L. Culverfoot.

Geum urbanum, L. Avens. Herb Bennet.

,, rivale, L. Water Avens.

Gladiolus communis, L. Gladiole. Corn-flag.

Glaucium luteum, L. Horned Poppy. Sea Poppy.

GLAUX MARITIMA, L. Black Saltwort. Sea Milk-wort.
GLYCERIA FLUITANS, RB. Manna-grass.
GNAPHALIUM DIOICUM, L. Cat's-foot.
,, MARGARITACEUM, L. Everlasting.
,, LUTEO-ALBUM, L. Jersey Livelong.
,, ULIGINOSUM, L. Cudweed. Chafe-weed. Cotton-weed.
,, GERMANICUM, L. Herb Impious. Childing Cudweed.
GOODYERA REPENS, RB. Creeping Satyrion.
GYROPHORA VELLEA, Ach. Rock-tripe.

HABENARIA BIFOLIA, RB. Butterfly Orchis.
,, VIRIDIS, RB. Frog Orchis.
HEDERA HELIX, L. Ivy.
HEDYSARUM CORONARIUM, L. French Honeysuckle.
HELIANTHEMUM VULGARE, L. Rock-rose. Sunflower.
HELIANTHUS ANNUUS, L. Sunflower.
,, TUBEROSUS, L. Jerusalem artichoke.
HELLEBORUS NIGER, L. Christmas Rose.
,, FŒTIDUS, L. Bear's foot. Oxheel. Setterwort.
,, VIRIDIS, L. Peg-roots.
HELMINTHIA ECHIOIDES, Gärt. Lang-de-beef. Oxtongue.
HERACLEUM SPHONDYLIUM, L. Cow Parsnip. Meadow Parsnip.
Madnep. Hog-weed. Clog-weed. Bear's breech.
HERMINIUM MONORCHIS, RB. Musk Orchis.
HERNIARIA GLABRA, L. Rupture-wort. Burst-wort.
HESPERIS MATRONALIS, L. Dame's Violet. Damask Violet.
Queen's Violet. Close Sciences.
,, TRISTIS, L. Melancholy Gentleman.
HIERACIUM, L. Hawk-bit. Hawk-weed.
,, PILOSELLA, L. Mouse-ear.
,, AURANTIACUM, L. Grimm the Collier.
HIEROCHLOE BOREALIS, Rm. Holy-grass.
HIPPOCREPIS COMOSA, L. Horse-shoe-vetch. Unshoe-the-horse.
HIPPOPHAE RHAMNOIDES, L. Sea Buck-thorn. Sallow-thorn.
Willow-thorn.
HIPPURIS VULGARIS, L. Mare's tail.
HORDEUM, L. Barley. Big. Bear.
,, MARITIMUM, L. Squirrel-tail.

HORDEUM MURINUM, L. Mouse Barley. Wall Barley.
,, PRATENSE, L. Rie-grass.
HOTTONIA PALUSTRIS, L. Water Violet. Bog Featherfoil. Water
Milfoil. Water Gilliflower.
HUMULUS LUPULUS, L. Hop.
HYACINTHUS, L. Hyacinth. Jacinth.
HYDROCHARIS MORSUS RANÆ, L. Frogbit.
HYDROCOTYLE VULGARIS, L. Penny Rot. White Rot. Flook-
wort. Marsh Pennywort.
HYMENOPHYLLUM, Sm. Filmy Fern.
HYOSCYAMUS NIGER, L. Henbane.
HYOSERIS MINIMA, L. Swine Succory.
HYPERICUM ANDROSÆMUM, L. Tutsan. Park-leaves.
,, PERFORATUM, L. St. John's wort.
,, QUADRANGULUM, L. St. Peter's wort. Hard-hay.
HYPOCHERIS MACULATA, L. Cat's ear.
HYSSOPUS OFFICINALIS, L. Hyssop.

IBERIS UMBELLATA, L. Candy-tuft.
,, AMARA, L. Sciatica Cress.
ILEX AQUIFOLIUM, L. Holly. Holm. Hulst. Hulver. Christmas.
IMPATIENS NOLI ME TANGERE, L. Touch-me-not. Balsamine.
Quick-in-Hand.
IMPERATORIA OSTRUTHIUM, L. Master-wort.
INULA CONYZA, DC. Ploughman's Spikenard. Fleawort.
,, DYSENTERICA, L. Fleabane Mullet.
,, HELENIUM, L. Elecampane. Elf-dock. Horse-hele.
Scab-wort.
,, CRITHMOIDES, L. Golden Samphire.
,, PULICARIA, L. Fleabane. Herb Christopher.
IRIS PSEUDACORUS, L. Sword Flag. Yellow Flag.
,, FLORENTINA, L. Orrice-root. Flower-de-Luce.
,, FŒTIDISSIMA, L. Stinking Gladdon. Gladwyn. Roastbeef.
ISATIS TINCTORIA, L. Woad.
ISOETES LACUSTRIS, L. Quill-wort.

JASIONE MONTANA, L. Sheep's-bit Scabious.
JASMINUM OFFICINALE, L. Jessamine.

JUGLANS REGIA, L. Walnut. French-nut.
JUNCUS, L. Rush. Seaves.
,, EFFUSUS, L. Candle Rush. Pin Rush.
,, SQUARROSUS, L. Goose-corn. Moss Rush.
JUNIPERUS COMMUNIS, L. Juniper.
,, SABINA, L. Savine.

LACTUCA SATIVA, L. Lettuce.
,, VIROSA, L. Sleepwort.
LAGURUS OVATUS, L. Hare's tail.
LAMINARIA SACCHARINA, Lam. Honeyware. Sea Belt.
,, DIGITATA, Ag. Tangle.
LAMIUM AMPLEXICAULE, L. Henbit.
,, PURPUREUM, L. Red Dead-nettle.
,, ALBUM, L. White Dead-nettle. Deaf-, or Blind-nettle.
White Archangel.
,, GALEOBDOLON, Crz. Yellow Archangel. Weasel-snout.
Yellow Dead-nettle.
LAPSANA COMMUNIS, L. Nipple-wort.
LATHRÆA SQUAMARIA, L. Toothwort. Clown's Lungwort.
LATHYRUS NISSOLIA, L. Grass Vetch.
,, APHACA, L. Yellow Vetchling.
,, MACRORRHIZUS, Wim. Heath Pea.
,, PRATENSIS, L. Tare-everlasting. Vetchling.
,, SYLVESTRIS, L. Everlasting Pea.
,, ODORATUS, L. Sweet Pea.
LAURUS NOBILIS, L. Sweet Bay. Roman Laurel. Lorer.
LAVATERA ARBOREA, L. Velvet-leaf. Tree Mallow.
LAVANDULA SPICA, L. Lavender.
,, STOECHAS, L. Cassidony. Stick-a-dove. French Lavender.
LECANORA TARTAREA, Ach. Cudbear.
LECIDEA GEOGRAPHICA, Hk. Map Lichen.
LEDUM PALUSTRE, L. Marsh Cistus.
LEMNA, L. Frog-foot. Duck's meat.
LEONTODON TARAXACUM, L. See TARAXACUM OFFICINALE, W.
LEONURUS CARDIACA, L. Motherwort.
LEPIDIUM CAMPESTRE, RB. Mithridate Pepperwort. Cow-cress.
,, RUDERALE, L. Bowyer's Mustard.

LEPIDIUM LATIFOLIUM, L. Penny Cress. Dittany. Dittander. Pepper-wort.

,, SATIVUM, L. Town Cress. Garden Cress.

LEUCOIUM ÆSTIVUM, L. Snowflake.

LIGUSTICUM SCOTICUM, L. Lovage.

LIGUSTRUM VULGARE, L. Privet. Primprint.

LILIUM CANDIDUM, L. White Lily. Juno's Rose.

LIMNANTHEMUM NYMPHÆOIDES, Link. Fringed Buckbean. Marsh flower.

LIMOSELLA AQUATICA, L. Mudwort.

LINARIA VULGARIS, Much. Toadflax. Flaxweed.

,, CYMBALARIA, Mill. Mother-of-thousands. Penny-wort.

,, SPURIA, S. Fluellin.

LINOSYRIS VULGARIS, Cas. Goldilocks.

LINUM USITATISSIMUM, L. Flax. Line. Linseed.

,, CATHARTICUM, L. Fairy-flax. Mill-mountain.

LISTERA OVATA, R.B. ·Twayblade. Bifoil.

LITHOSPERMUM OFFICINALE, L. Gromwell. Pearl-plant. Lichwale.

,, ARVENSE, L. Bastard Alkanet.

LITTORELLA LACUSTRIS, L. Shoreweed.

LOLIUM PERENNE, L. Ray-grass.

,, TEMULENTUM, L. Darnel. Ivray. Ray. Neele.

LONICERA CAPRIFOLIUM, L. Caprifoly.· Honeysuckle. Lily among thorns.

,, PERICLYMENUM, L. Woodbine.

,, XYLOSTEUM, L. Fly Honeysuckle.

LOTUS CORNICULATUS, L. Bird's-foot Trefoil. Butter-jags. Crowtoes.

LUNARIA BIENNIS, L. Honesty. Money-flower. Satin-flower. Bolbonac.

LUPINUS, L. Lupine.

LUZULA SYLVATICA, B. Woodrush.

,, CAMPESTRIS, B. Cuckoo-grass.

LYCHNIS GITHAGO, Lam. Corn-cockle. Gith. Corn Campion. Corn Pink.

,, FLOS CUCULI, L. Ragged Robin. Cuckoo-flower, or -Gilliflower. Meadow Pink. Meadow Campion.

Lychnis Viscaria, L. Catch-fly.

,, Chalcedonica, L. Bristol Flower. Cross of Jerusalem.
Flower of Constantinople.

,, coronaria, L. Rose Campion.

Lycoperdon, L. Fist-ball. Puck-fist. Fuss-ball. Puff-ball.

Lycopodium clavatum, L. Wolf's claw. Club-moss.

,, alpinum, L. Heath Cypress.

,, Selago, L. Fir-moss.

Lycopsis arvensis, L. Wild Bugloss.

Lycopus europæus, L. Gipsy-wort. Water Horehound.

Lysimachia vulgaris, L. Loosestrife.

,, nemorum, L. Yellow Pimpernell.

,, nummularia, L. Herb Twopence. Moneywort. Twopenny-grass.

Lythrum hyssopifolium, L. Grass Poley.

,, Salicaria, L. Purple Loosestrife.

Maianthemum bifolium, DC. One-blade.

Malva rotundifolia, L. Dwarf Mallow.

,, sylvestris, L. Mallow. Round Dock. Maule. Hock-
herb.

,, moschata, L. Musk Mallow.

Marchantia polymorpha, L. Liverwort.

Marrubium vulgare, L. Horehound.

Matricaria Chamomilla, L. Dog's Chamomile. Maithe. Mather.
Maudlin. Mayweed.

Matthiola incana, RB. Stock-Gilliflower. July flower. Stock.

Meconopsis cambrica, V. Welsh Poppy.

Medicago sativa, L. Lucern. Medick. Snail Clover. Sainfoin. Holy hay.

,, lupulina, L. Nonsuch. Shamrock. Black-seed.

,, maculata, Willd. Heart Clover.

Melampyrum pratense, L. Cow-wheat.

,, sylvaticum, L. Horse-flower.

Melilotus officinalis, L. Melilot. Hart's Clover. King's
Clover. Plaister Clover.

Melissa officinalis, L. Balm.

Melittis Melissophyllum, L. Bastard Balm.

MENTHA SYLVESTRIS, L. Horse Mint. Brook-, or Water-Mint.
   ,,   VIRIDIS, L. Spear Mint. Garden Mint.
   ,,   PIPERITA, L. Pepper Mint.
   ,,   CITRATA, Ehr. Bergamot Mint.
   ,,   PULEGIUM, L. Pennyroyal. Pudding-grass.
   ,,   SATIVA, L. Garden Mint.
MENYANTHES TRIFOLIATA, L. Buckbean. Bogbean. Marsh Trefoil.
MENZIESIA POLIFOLIA, Sm. St. Daboec's Heath. Irish Heath.
MERCURIALIS PERENNIS, L. Dog's Mercury.
   ,,   ANNUA, L. French Mercury.
MERULIUS LACRYMANS, Wulf. Dry-rot.
MESPILUS GERMANICA, L. Medlar.
MEUM ATHAMANTICUM, Jac. Baldmoney. Mew. Bearwort.
     Spikenel.
MILIUM EFFUSUM, L. Millet.
MONARDA FISTULOSA, L. Bergamot.
MONOTROPA HIPOPITYS, L. Yellow Bird's-nest. Pine-sap.
MONTIA FONTANA, L. Water-Blinks, or -Chickweed. Blinking
     Chickweed.
MORCHELLA ESCULENTA, P. Morel.
MORUS NIGRA, W. Mulberry.
MUSCARI RACEMOSUM, M. Grape Hyacinth.
MYOSOTIS PALUSTRIS, With. Forget-me-not. Mouse-ear Scorpion-
     grass.
MYOSURUS MINIMUS, L. Mousetail. Blood-strange.
MYRICA GALE, L. Dutch Myrtle. Gale. Bog Myrtle. Sweet
     Willow. Candleberry.
MYRIOPHYLLUM, L. Water Milfoil.
MYRRHIS ODORATA, Scop. Sweet Chervil. Sweet Cicely.
MYRTUS COMMUNIS, L. Myrtle.

NARCISSUS PSEUDONARCISSUS, L. Daffodil. Lent-lily. Crow-
     bells. Affadyl.
   ,,   BIFLORUS, Curt. Primrose Peerless.
   ,,   JONQUILLA, L. Jonquil.
NARDUS STRICTA, L. Matgrass. Wirebent.
NARTHECIUM OSSIFRAGUM, Huds. Bog or Marsh Asphodel. Lan-
     cashire Asphodel.

NASTURTIUM OFFICINALE, L. Water-cress.

NEOTTIA NIDUS AVIS, L. Bird's nest.

NEPETA GLECHOMA, Benth. Ground Ivy. Alehoof. Cat's foot. Hove. Gill. Haymaids. Tunhoof.

   ,,   CATARIA, L. Cat Mint. Nep.

NERINE SARNIENSIS, W. Guernsey Lily.

NIGELLA DAMASCENA, L. Devil-in-a-bush. Catharine's flower. Love-in-a-mist. Bishop's wort. Fennel-flower. Kiss me twice.

NUPHAR LUTEUM, Sm. Yellow Water-lily. Brandy-bottle. Candock.

NYMPHÆA ALBA, L. White Water-lily.

OCYMUM BASILICUM, L. Sweet Basil.

ŒNANTHE FISTULOSA, L. Water Dropwort.

   ,,   CROCATA, L. Water Hemlock. Dead-tongue.

   ,,   PHELLANDRIUM, Lam. Horse-bane.

ŒNOTHERA BIENNIS, L. Evening Primrose.

OMPHALODES VERNA, Mn. Venus' Navel-wort.

ONOBRYCHIS SATIVA, L. Sainfoin. French-grass. Cock's comb.

ONONIS ARVENSIS, L. Rest-harrow. Ground Furze. Stay-plough.

ONOPORDON ACANTHIUM, L. Scotch Thistle. Cotton Thistle. Silver Thistle.

OPHIOGLOSSUM VULGATUM, L. Adder's tongue.

OPHRYS APIFERA, Huds. Bee Orchis.

   ,,   ARANIFERA, Huds. Spider Orchis.

   ,,   FUCIFERA, Sm. Drone Orchis.

   ,,   MUSCIFERA, Huds. Fly Orchis.

ORCHIS MORIO, L. Goose and Goslings. Gandergosses.

   ,,   MILITARIS, L. Soldier Orchis.

   ,,   MASCULA, L. Long purples. Stander-wort.

   ,,   MACULATA, L. Dead-man's-fingers. Hand Orchis.

   ,,   HIRCINA, Scop. Lizard Orchis.

   ,,   LATIFOLIA, L. Salep.

ORIGANUM VULGARE, L. Marjoram. Organy.

ORNITHOGALUM PYRENAICUM, L. French Sparrow-grass.

   ,,   UMBELLATUM, L. Star of Bethlehem. Eleven-o'clock-lady.

ORNITHOPUS PERPUSILLUS, L. Bird's foot.

OROBANCHE MAJOR, L. Broomrape.

OROBUS, see VICIA.

OSMUNDA REGALIS, L. Osmund-royal. Flowering-fern. Herb
    Christopher.

OXALIS ACETOSELLA, L. Wood Sorrel. Hallelujah. Gowk-meat.
    Cuckoo's bread. Cuckoo Sorrel. Wood Sower.
    Stub-wort. Stab-wort.

OXYRIA RENIFORMIS, Sm. Mountain Sorrel.

PÆONIA CORALLINA, L. Peony. Marmaritin. Chesses.

   ,, MOUTAN, L. Moutan.

PALMELLA CRUENTA, Agh. Gory Dew.

PANICUM CRUS GALLI, L. Cock's-spur-grass.

PAPAVER RHŒAS, L. Corn or Red Poppy. Corn Rose. Cop
    Rose. Red-weed. Canker Rose. Headache. Head-
    warke. Joan Silver-pin. Cheese-bouls.

   ,, SOMNIFERUM, L. White Poppy. Cheesebowls. Mawseed.

PARIETARIA OFFICINALIS, L. Pellitory-of-the-wall. Lichwort.

PARIS QAUDRIFOLIA, L. Herb Trulove. One-berry. Herb Paris.
    Four-leaved-grass. Leopard's bane.

PARNASSIA PALUSTRIS, L. Grass of Parnassus.

PASTINACA SATIVA, L. Parsnip.

PEDICULARIS SYLVATICA, L. Lousewort. Red Rattle.

PEPLIS PORTULA, L. Water Purslane.

PETROSELINUM SATIVUM, L. Parsley.

   ,, SEGETUM, L. Corn Hone-wort.

PEUCEDANUM OFFICINALE, L. Brimstone-wort. Sulphur-wort.
    Hog's Fennel. Harstrong. Spreusidany. Cammock.

   ,, PALUSTRE, Mn. Milk Parsley.

PHALARIS CANARIENSIS, L. Bird-seed. Canary-grass.

PHALLUS IMPUDICUS, L. Stinkhorn.

PHASCUM, L. Earth Moss.

PHASEOLUS VULGARIS, L. French Bean. Kidney Bean.

   ,, MULTIFLORUS, W. Scarlet runner.

PHILADELPHUS CORONARIA, L. Syringa.

PHLEUM PRATENSE, L. Timothy-grass.

PHLOMIS FRUTICOSA, L. Jerusalem Sage.

PHYSALIS ALKEKENGI, L. Winter Cherry.

PILULARIA GLOBULIFERA, L. Pill-wort. Pepper-grass.

PIMPINELLA SAXIFRAGA, L. Burnet Saxifrage. Pimpinell.

   ,,   ANISUM, L. Anise.

PINGUICULA VULGARIS, L. Butterwort. Rot-grass. Yorkshire
     Sanicle. Bog Violet.

PINUS SYLVESTRIS, L. Scotch Fir. Riga Pine.

   ,,   ABIES, L. Spruce Fir.

   ,,   PICEA, L. Silver Fir.

PISUM SATIVUM, W. Pea.

   ,,   MARITIMUM, L. Sea Pea.

PLANTAGO LANCEOLATA, L. Ribwort Plantain. Kemps. Cocks.

   ,,   MEDIA, L. Lamb's tongue.

   ,,   CORONOPUS, L. Star of the Earth. Hart's horn.

   ,,   MAJOR, L. Plantain. Way-bread.

PLATANUS, L. Plane. Platane.

POA FLUITANS, Scop. Flote-grass.

POLEMONIUM CÆRULEUM, L. Jacob's ladder. Makebate. Greek
     Valerian.

POLYCARPON TETRAPHYLLUM, L. Allseed.

POLYGALA VULGARIS, L. Rogation flower. Gang flower. Pro-
     cession flower. Milk-wort. Cross flower.

POLYGONATUM MULTIFLORUM, All. Solomon's seal. Lady's seal.

POLYGONUM AVICULARE, L. Knotgrass. Ninety-knot. Swine's
     grass. Sparrow-tongue. Bird's tongue. Pink-weed.
     Centinode.

   ,,   FAGOPYRUM, L. Buckwheat. Brank.

   ,,   CONVOLVULUS, L. Black Bindweed.

   ,,   BISTORTA, L. Bistort. Snake-weed. Twice-writhen.
     Adder-wort. Red-legs. Osterick.

   ,,   PERSICARIA, L. Persicaria. Peach-wort. Red-shanks.

   ,,   HYDROPIPER, L. Arsmart. Culrage. Ciderage. Water
     Pepper. Lake-weed. Red-knees.

POLYPODIUM VULGARE, L. Common Polypody. Wall-fern. Oak-
     fern.

   ,,   PHEGOPTERIS, L. Beech-fern.

   ,,   DRYOPTERIS, L. Oak-fern.

POLYPORUS IGNIARIUS, L. Amadou. Touch-wood.

POPULUS ALBA, L. Abele. White Poplar.

,, TREMULA, L. Aspen.

,, NIGRA, L. Black Poplar.

,, ,, var. FASTIGIATA, Dsf. Lombardy Poplar.

PORPHYRA LACINIATA, Ag. Laver.

PORTULACA OLERACEA, L. Purslane.

POTAMOGETON, L. Pond-weed.

,, DENSUS, L. Frog's Lettuce.

,, NATANS, L. Tench-weed.

POTENTILLA REPTANS, L. Cinquefoil. Five-leaf. Five-finger-grass.

,, TORMENTILLA, L. Tormentil. Blood-root. Septfoil.

,, FRAGARIASTRUM, Ehr. Barren Strawberry.

,, ANSERINA, L. Silver-weed. Goose Tansey. Goose-grass.

,, COMARUM, L. Purple-wort.

POTERIUM SANGUISORBA, L. Burnet. Pimpinell. Salad Burnet.

PRENANTHES MURALIS, L. Wall Lettuce.

PRIMULA VERIS, L. a. Primrose. Petty Mullein. Primerole.

,, ,, b. Cowslip. Herb Peter. Paigle. Palsy-wort.

,, ,, c. Oxlip. Polyanthus.

,, AURICULA, L. Bear's ears. Mountain Cowslip.

,, FARINOSA, L. Bird's eye.

PRUNELLA VULGARIS, L. Hook-heal. Self-heal. Carpenter's herb. Sickle-wort.

PRUNUS COMMUNIS, Huds. v. spinosa, Sloe. Blackthorn. Skeg. Snag.

,, ,, v. insititia, Bullace. Damson. Plum.

,, AVIUM, L. Gean. Mazzards. Merry.

,, ARMENIACA, L. Apricot.

,, PADUS, L. Bird Cherry. Heg-, or Hack-berry.

,, LAUROCERASUS, L. Laurel. Cherry Laurel.

,, LUSITANICA, L. Portugal Laurel.

PSAMMA ARENARIA, P.B. Maram. Halm. Matweed. Stare.

PTERIS AQUILINA, L. Brake. Bracken.

PULMONARIA OFFICINALIS, L. Lungwort. Bugloss Cowslip. Jerusalem Cowslip.

PYRETHRUM PARTHENIUM, L. Feverfew. Bertram. Mayweed. Maithes.

PYROLA, L. Wintergreen.
,,   SECUNDA, L. Yevering Bells.
PYRUS COMMUNIS, L. Pear. Choke Pear.
,,   MALUS, L. Apple. Crab. Codlin. Wharre.
,,   ARIA, Sm. White Beam.
,,   TORMINALIS, Sm. Sorb. Wild Service. Swallow
      Pear.
,,   AUCUPARIA, G. Rowan. Mountain Ash.
,,   DOMESTICA, Sm. Service. Chequer-tree.

QUERCUS ROBUR, L. Oak.
,,   ILEX, L. Holm Oak. Evergreen Oak.
,,   CERRIS, L. Turkey Oak.
,,   ,,   var. Luccombe Oak.

RADIOLA MILLEGRANA, L. All-seed. Flax-seed.
RANUNCULUS ACONITIFOLIUS, L. Fair Maids of France.
,,   AQUATICUS, L. Water Crowfoot.
,,   BULBOSUS, L. St. Anthony's Rape, or Turnip.
,,   LINGUA, L. Spear-wort.
,,   FLAMMULA, L. Bane-wort.
,,   FICARIA, L. Pile-wort. Lesser Celandine.
,,   SCELERATUS, L. Celery-leaved Crowfoot.
,,   AURICOMUS, L. Goldilocks.
,,   ACRIS, L. Crowfoot. Buttercup. Gold-cup. Gold-
      knoppes. King's-cup. Crazy.
,,   ARVENSIS, L. Hunger-weed.
RAPHANUS RAPHANISTRUM, L. Jointed Charlock. Runch. Ra-
      bone.
,,   SATIVUS, L. Radish.
RESEDA LUTEA, L. Base Rocket.
,,   LUTEOLA, L. Weld. Yellow-weed. Dyer's Rocket.
,,   ODORATA, L. Mignonette.
RHAMNUS CATHARTICUS, L. Buckthorn. Hart's-thorn. Rhine-
      berries. Rain-berry thorn.
,,   FRANGULA, L. Black Alder. Berry Alder. Butcher's
      Prick-wood.
,,   ALATERNUS, L. Barren Privet.

RHEUM, L. Rhubarb.

RHINANTHUS CRISTA GALLI, L. Yellow Rattle. Penny-grass. Rattle-box.

RHODOMENIA PALMATA, B.V. Dulse.

RIBES GROSSULARIA, L. Gooseberry. Thape. Feabe.

  ,,   RUBRUM, L. Currant. Garnet-berry. Raisin-tree.

  ,,   NIGRUM, L. Black Currant. Squinancy berry. Gazles.

ROBINIA PSEUDACACIA, L. Acacia.

ROCCELLA TINCTORIA, L. Archal. Orchil. Litmus. Cork.

ROSA PIMPINELLIFOLIA, L. Burnet Rose.

  ,,   RUBIGINOSA, L. Sweet-briar. Eglantine.

  ,,   CANINA, L. Dog Rose. Hip Rose.

ROTTBOELLIA INCURVATA, L. Hard-grass.

RUBIA TINCTORUM, W. Madder. Warence.

RUBUS CÆSIUS, L. Dewberry. Theve-thorn.

  ,,   IDÆUS, L. Raspberry. Framboise. Hindberry.

  ,,   FRUTICOSUS, L. Bramble. Blackberry. Bumble-kite. Scald-berry.

  ,,   CHAMÆMORUS, L. Cloud-berry. Noops.

  ,,   SAXATILIS, L. Knot-berry.

RUMEX, L. Dock.

  ,,   SANGUINEUS, L. Blood-wort. Bloody Dock.

  ,,   PULCHER, L. Fiddle Dock.

  ,,   ACETOSA, L. Sorrel. Sharp Dock. Green-sauce.

  ,,   ACETOSELLA, L. Sheep Sorrel.

  ,,   OBTUSIFOLIUS, L. Butter Dock.

  ,,   PATIENTIA, L. Monk's Rhubarb.

RUPPIA MARITIMA, L. Sea-grass. Tassel-grass.

RUSCUS ACULEATUS, L. Butcher's broom. Box Holly. Knee Holly. Pettigrew.

  ,,   RACEMOSUS, L. Alexandrian Laurel.

RUTA GRAVEOLENS, L. Rue. Herb of Grace.

SAGINA PROCUMBENS, L. Pearl-wort.

SAGITTARIA SAGITTIFOLIA, L. Arrow-head,

SALICORNIA HERBACEA, L. Glass-wort. Frog-grass. Sea Grape. Salt-wort. Marsh Samphire. Crab-grass.

SALIX, Willow.

"   FRAGILIS, L.   Crack Willow.

"   ALBA, L.   White Willow.

"   VIMINALIS, L.   Osier. Withy.

"   CAPREA, L.   Sallow. Palm.

"   HERBACEA, L.   Dwarf Willow.

"   BABYLONICA, L.   Weeping Willow.

SALSOLA KALI, L.   Sowd-wort.   Prickly Glass-wort.

SALVIA OFFICINALIS, L.   Sage.

"   PRATENSIS, L.   Meadow Clary.

"   VERBENACA, L.   Wild Clary.

"   SCLAREA, L.   Clary. Sebright. Clear-eye.

SAMBUCUS EBULUS, L.   Dwarf Elder.   Dane-wort.   Wall-wort.

"   NIGRA, L.   Elder.   Bore-tree.

SAMOLUS VALERANDI, L.   Brook-weed.   Water Pimpernel.

SANGUISORBA OFFICINALIS, L.   Great Burnet.   Burnet Blood-
wort.

SANICULA EUROPÆA, L.   Sanicle.   Self-heal.

SAPONARIA OFFICINALIS, L.   Soap-wort.   Fuller's herb.   Bruisewort.

SARGASSUM VULGARE, Ag.   Gulf-weed.

SAROTHAMNUS SCOPARIUS, Wim.   Broom.

SATUREJA HORTENSIS, L.   Savoury.

SAXIFRAGA, L.   Saxifrage.   Thirlstane.   Stonebreak.

"   TRIDACTYLITES, L.   Nail-wort.   Whitlow-grass.

"   UMBROSA, L.   London Pride.   St. Patrick's-cabbage.
Pratling Parnel.   None-so-pretty.   Nancy-pretty.

SCABIOSA, L.   Scabious.

"   SUCCISA, L.   Devil's-bit.   Forebitten More.   Blue-
caps.

"   COLUMBARIA, L.   Small Scabious.

"   ARVENSIS, L.   Blue-caps.

SCANDIX PECTEN, L.   Venus' comb.   Shepherd's needle.   Devil's
darning-needles.   Lady's comb.   Crow-, or Crake-
needles.

SCILLA VERNA, Huds.   Star Hyacinth.

"   NUTANS, Sm.   Bluebell.   Harebell.   Squill.   Crowleeks.
Culverkeys.

SCIRPUS CÆSPITOSUS, L. Deer's hair.

,, LACUSTRIS, L. Bulrush.

SCLERANTHUS ANNUUS, L. Knawel.

SCOLOPENDRIUM VULGARE, Gärt. Hart's tongue.

SCORZONERA EDULIS, Mn. Viper-grass.

SCROPHULARIA NODOSA, L. Kernel-wort. Bull-wort. Brownwort.

,, AQUATICA, L. Bishop's leaves.

SCUTELLARIA GALERICULATA, L. Skull-cap.

,, MINOR, D. Hedge Hyssop.

SCYPHOPHORUS PYXIDATUS, Hk. Cup Lichen. Cup Moss.

SECALE CERFALE, L. Rye.

SEDUM RHODIOLA, DC. Rose-root.

,, TELEPHIUM, L. Orpine. Livelong. Liblong. Mid-
summer-men.

,, ALBUM, L. Worm-grass.

,, ACRE, L. Wall Pepper. Bird's bread. Jack-of-the-
buttery. Pricket. Pepper-crop.

,, REFLEXUM, L. Stonor. Trip-madam.

SEMPERVIVUM TECTORUM, L. Houseleek. Jupiter's beard. Joubarb. Ayegreen.

SENEBIERA CORONOPUS, Poir. Swine Cress. Wart Cress.

SENECIO VULGARIS, L. Groundsel. Simson.

,, JACOBÆA, L. Ragwort. Seggrum. Stagger-wort.
Staver-wort.

,, SARACENICUS, L. Saracen's Consound.

SERRATULA TINCTORIA, L. Saw-wort.

SESLERIA CÆRULEA, Scop. Moor-grass.

SHERARDIA ARVENSIS, L. Field Madder. Spur-wort.

SIBBALDIA PROCUMBENS, L. Scotch Cinquefoil.

SIBTHORPIA EUROPÆA, L. Cornish Money-wort. Penny-wort.

SILAUS PRATENSIS, B. Meadow Saxifrage.

SILENE ACAULIS, L. Moss Campion. Cushion Pink.

,, INFLATA, L. Ben. Bladder Campion. Knap-bottle.
Spatling Poppy.

,, MARITIMA, L. Witches' thimble.

SINAPIS, L. See BRASSICA.

SISON AMOMUM, L. Stone Parsley. Hone-wort.

SISYMBRIUM OFFICINALE, Scop. Hedge Mustard. Bank Cress. Crambling Rocket.

,, IRIO, L. London Rocket.

,, SOPHIA, L. Flixweed.

SIUM LATIFOLIUM, L. Water Parsnip.

,, SISARUM, L. Skirret.

SMYRNIUM OLUS ATRUM, L. Alexanders. Horse Parsley. Stanmarch.

SOLANUM DULCAMARA, L. Bittersweet. Woody Nightshade. Blue Bindweed. Felon-wort.

,, NIGRUM, L. Petty Morel.

,, TUBEROSUM, L. Potato.

,, LYCOPERSICUM, L. Gold apples. Love apple. Tomato.

,, MELONGENA, W. Egg-plant. Bringal.

SOLIDAGO VIRGA AUREA, L. Golden-rod.

SONCHUS OLERACEUS, L. Sow Thistle. Hare's Lettuce.

SPARGANIUM RAMOSUM, L. Bur Reed. Bede Sedge.

SPARTINA STRICTA, Sm. Cord-grass. Spart-, or Spurt-grass.

SPERGULA ARVENSIS, L. Spurry. Franke. Yarr.

SPERGULARIA RUBRA, Pers. Sand Spurry.

SPHAGNUM, L. Bog Moss.

SPIRÆA ULMARIA, L. Meadow-sweet. Mead-wort. Queen of the mead. Bride-wort.

,, FILIPENDULA, L. Drop-wort.

SPIRANTHES AUTUMNALIS, Rich. Lady's tresses.

STACHYS BETONICA, Benth. Betony.

,, GERMANICA, L. Wound-wort.

,, SYLVATICA, L. Red Archangel. Hedge Dead-nettle.

,, PALUSTRIS, L. Clown's Allheal.

STAPHYLEA PINNATA, L. Bladder-nut.

STATICE LIMONIUM, L. Sea Lavender.

,, ARMERIA, Sm. See ARMERIA.

STELLARIA MEDIA, L. Chickweed.

,, HOLOSTEA, L. All-bone. Stitch-wort.

STICTA PULMONARIA, Hk. Tree Lungwort.

STIPA PENNATA, L. Feather-grass.

STRATIOTES ALOIDES, L. Water-soldier. Knight's wort.

SUBULARIA AQUATICA, L. Awl-wort.

Symphoria racemosa, Ph. Snow-berry bush.
Symphytum officinale, L. Comfrey. Knit-back.
Syringa vulgaris, L. Lilac. Pipe-tree. Syring.

Tamarix gallica, L. Tamarisk.
Tamus communis, L. Black Bryony. Lady's seal.
Tanacetum vulgare, L. Tansy.
Taraxacum officinale, W. Dandelion. Priest's crown. Swine's
    snout. Pissabed. Blow-ball.
Taxus baccata, L. Yew. Palm.
Teesdalia nudicaulis, R.B. Shepherd's Cress.
Teucrium Scorodonia, L. Wood Sage. Wood Germander. Hind-
    heal.
   ,,   Scordium, L. Water Germander.
   ,,   Chamædrys, L. Germander. Horsechire.
   ,,   Botrys, L. Oak of Jerusalem.
Thalictrum flavum, L. Meadow Rue. Fen Rue.
Thesium linophyllum, L. Bastard Toadflax.
Thlaspi arvense, L. Penny Cress. Mithridate Mustard.
Thymus Serpyllum, L. Thyme. Hill Thyme. Pell-a-mountain.
    Mother of Thyme.
   ,,   vulgaris, L. Garden Thyme.
Tilia europæa, L. Lime-tree. Linden. Teil-tree. Bast-tree.
Tofieldia palustris, Huds. Scottish Asphodel.
Tordylium maximum, L. Hart-wort.
Tragopogon pratense, L. Goat's beard. Go-to-bed-at-noon.
    Noon-flower. Noontide. Buck's beard. Joseph's
    flower.
   ,,   porrifolium, L. Salsify. Star of Jerusalem. Nap-
    at-noon.
Trapa natans, L. Saligot.
Tremella Nostoc, L. Fallen Stars. Nostoc. Will of the
    Wisp.
   ,,   arborea, Sm. Fairy butter.
Trichomanes radicans, Sw. Bristle-fern.
Trientalis europæa, L. Winter-green Chickweed.
Trifolium arvense, L. Haresfoot Clover.
   ,,   pratense, L. Meadow Clover. Honeysuckle.

TRIFOLIUM HYBRIDUM, L. Alsike.
,, FRAGIFERUM, L. Strawberry Clover.
,, INCARNATUM, L. Crimson Clover.
,, AGRARIUM, L. Hop Clover.
,, REPENS, L. Dutch Clover.
TRIGLOCHIN PALUSTRE, L. Arrow-grass.
TRIGONELLA ORNITHOPODIOIDES, DC. Bird's foot.
TRINIA GLABERRIMA, L. Hone-wort.
TRITICUM REPENS, L. Couch. Quitch grass.
,, CANINUM, Huds. Dog grass.
,, SPELTA, L. Spelt. Starch corn.
TROLLIUS EUROPÆUS, L. Globe flower. Troll flower. Locken
    Gowan.
TROPÆOLUM MAJUS, L. Sturtion.
TUBER CIBARIUM, Sib. Earth-ball. Truffle.
TULIPA, L. Tulip. Turk's cap.
TUSSILAGO PETASITES, L. Butterbur Coltsfoot. Pestilence
    weed.
,, FARFARA, L. Coltsfoot. Asses foot. Bull's foot. Fole-
    foot. Cough-wort. Horse-hoof.
TYPHA LATIFOLIA, L. Reed-mace. Bull-segg. Dunse-down.
    Water torch. Marsh beetle. Club Rush.

ULEX EUROPÆUS, L. Furze. Thorn Broom.
ULMUS CAMPESTRIS, Sm. Elm-tree.
,, MONTANA, Sm. Wych Elm. Wych Hazel.
ULVA LATISSIMA, Grev. Laver.
UMBILICUS, DC. See COTYLEDON.
UREDO CARIES, L. Smut.
URTICA DIOICA, L. Nettle.
,, URENS, L. Sting Nettle.
,, PILULIFERA, L. Roman Nettle.
UTRICULARIA VULGARIS, L. Bladder-wort. Hooded Milfoil.

VACCINIUM MYRTILLUS, L. Bilberry. Whortle-berry. Hurtle-
    berry. Wim-berry. Black-worts.
,, ULIGINOSUM, L. Blea-berry.

VACCINUM VITIS IDÆA, L. Cow-berry.

,, OXYOOCCOS, L. Cran-berry. Bog-berry. Bog-wort. Fen-berry. Marsh-worts.

VALERIANA OFFICINALIS, L. Valerian. Setwal. Capon's tail. Cut-heal. Herb Bennett.

VALERIANELLA OLITORIA, Poll. Lamb's Lettuce. Corn salad. White potherb.

VERBASCUM THAPSUS, L. Mullein. Hagtaper. Torch. Higtaper. Hare's beard. Jupiter's staff. Bullock's Lungwort. Velvet Dock.

,, BLATTARIA, L. Moth Mullein.

VERBENA OFFICINALIS, L. Vervain. Simpler's joy. Holy herb. Juno's tears. Pigeon's grass.

VERONICA OFFICINALIS, L. Ground-heele.

,, SERPYLLIFOLIA, L. Paul's Betony.

,, BECCABUNGA, L. Brooklime. Water Pimpernel.

,, CHAMÆDRYS, L. Speedwell. Blue Bird's-eye.

,, AGRESTIS, L. Germander Chickweed.

,, HEDERIFOLIA, L. Morgeline. Winter-weed.

VIBURNUM LANTANA, L. Wayfaring-tree. Whitten-tree. Lithytree. Mealy-tree.

,, OPULUS, L. Guelder Rose. Water Elder. Rose Elder. Snow-ball.

,, TINUS, L. Laurestinus.

VICIA SATIVA, L. Vetch. Fitch.

,, FABA, L. Horse Bean.

,, OROBUS, DC. Bitter Vetch. Kipper nut. Cormeille.

,, LATHYROIDES, L. Strangle-tare.

,, HIRSUTA, K. Tine-tare.

VINCA MAJOR, L. Periwinkle.

VIOLA ODORATA, L. Sweet Violet. March Violet.

,, CANINA, L. Dog Violet.

,, TRICOLOR, L. Pansy. Heartsease. Herb Trinity. Three faces under a hood. Fancy. Flamy. Kiss me. Love in idle.

VISCUM ALBUM, L. Mistletoe.

VITIS VINIFERA, W. Vine.

XANTHIUM STRUMARIUM, L.   Bur-weed.   Burdock Clot-weed.
Ditch-bur.   Louse-bur.

ZANNICHELLIA PALUSTRIS, L.   Horned Pond-weed.
ZOSTERA MARINA, L.   Grass-wrack.

THE END.

STEPHEN AUSTIN, PRINTER, HERTFORD.

www.ingramcontent.com/pod-product-compliance
Lightning Source LLC
Chambersburg PA
CBHW021503210326
41599CB00012B/1111